SPILL THE NFT - A BEGINNER'S GUIDE TO NON-FUNGIBLE TOKENS

HOW NFTS WORK, HOW THEY ARE CREATED AND HOW TO MAKE MONEY FROM THEM

MAX V PALMEIRA

MVP

CONTENTS

Introduction   vii

1. WELCOME TO THE WORLD OF NFTS   1
Understanding Blockchain Technology   4
Cryptocurrency and Decentralization   7
Economic Freedom   8
Protecting User Interests   9
Borderless Payments   9
Wider Financial Inclusion   9
Let's Talk About Fungibility   10
Is the NFT Space a Bubble?   11

2. A BRIEF HISTORY OF NFTS   15
The Colored Coins   15
Counterparty   17
Going Mainstream   19
Most Expensive NFTs   22
Beeple's Creations   22
Noora Health – Save Thousands of Lives   23
This Changed Everything   23
Art Blocks Collection – Ringers #109   24
Right-Click and Save As Guy - Xcopy   24
CryptoPunk – Larva Labs   25
Pak – The Merge   26
Most Famous NFTs   26
NBA TopShot   26
CryptoKitties   29
Jack Dorsey's First Tweet   30
Beeple   31

3. HOW THEY WORK - LOOKING UNDER
THE BONNET OF NFTS   33
The Case for Centralization   39
How the Technology Works   40
Hashing   42
Practical Blockchain Applications   42
Benefits of NFTs   45
Human Psychology and the Perception of
Value   47

4. THE DIFFERENT TYPES OF NFTS   49
Artwork   50
Collectibles and Trading Cards   51
Domain Names   52
Event Tickets   53
Gaming NFTs   54
Major Sporting Moments   55
Meme NFTs   56
Music and Media   57
Real Assets   58
Virtual Fashion   60

5. HOW TO MAKE MONEY FROM NFTS   62
Setting Up Your Account   63
Community Value in NFTs   64
Watch This Space   69

6. NFT MARKETPLACES   73
Choosing the Right Marketplace   75
Comparing Your Options   78
Open Marketplaces   79
OpenSea   79
Rarible   80
Hic Et Nunc   81
Mintable   82
Zora   83
Closed Marketplaces   84
Dartroom   84
Foundation   85

Makersplace   87

Nifty Gateway   88

SuperRare   88

7. HOW TO CREATE SUCCESSFUL NFT
ART WITH A THRIVING COMMUNITY   90

Building a Profitable NFT Brand   91

Creating Value in the Community   100

Leveraging Social Media   101

Building an NFT Community   103

8. CRYSTAL BALL TIME - THE FUTURE
OF NFTS   107

A Thriving Community   110

A Comparative Analysis   113

The Future of Finance   117

The Next Step   120

9. CONCLUSION   124

10. A QUICK FAVOR   128

11. REFERENCES   129

Max V. Palmeira

INTRODUCTION

From Aruba to Zambia, and everywhere in between, people are talking about NFTs. They are the new craze in town.

Funnily enough, though, NFTs are not that new. The current hype around them is because they have now well and truly hit the mainstream stage, and it is a time when their value and relevance in the world of blockchain technology is becoming increasingly important.

Many people learned about NFTs from celebrities dabbling in the subject. If you follow the likes of Lindsay Lohan and Snoop Dogg, you might already be in the know. However, there's more to NFTs than celebrity affiliation. NFTs are a unique game changer. Not only are they making headlines in celebrity spaces, but they are poised to play a crucial role in the next stage of the evolution of the internet—the metaverse. (Yup - the very phrase that Mark Zuckerberg keeps banging on about!)

This book will give you an understanding of NFTs in a simple and comprehensible way so that you, if that way inclined, can jump in on this growing trend.

When new technologies like NFTs come up, many people miss out on the amazing opportunities to learn and maximize on the investment opportunities that they bring.

This could be because of the notion that new technologies are only the purview of tech enthusiasts and programmers, and that anyone who doesn't fit into those won't stand to benefit. This isn't the case with NFTs—and hopefully this book will show you why.

To help you understand the dynamics of NFTs, this book will get you started with the basics of blockchain technology, on which NFTs are created.

You have probably come across this term before, especially when talking about Bitcoin and other cryptocurrencies. Blockchain is the foundation upon which these currencies exist. We'll discuss the unique features of blockchain technology which underpin all the cryptocurrencies and NFTs. This will also help you understand the important concept of decentralization, and how it's a welcome shift from our traditional understanding of money.

We will also talk about digital assets, an exciting category whose understanding can help you identify worthwhile potential investments for the future.

Investments, for the longest time, have been limited to traditional investment avenues in the financial markets.

This includes bonds, stocks, debentures, real estate, commodities, foreign exchange, and so on.

With digital assets, you have even more access and options when it comes to investing. Take Bitcoin, for example. At the time of this publication, Bitcoin was exchanging at around $30,700 to the US Dollar. Yet, there was a time when the value of a Bitcoin was so insignificant, it was exchanging at well below $1.

Now, imagine you bought a thousand Bitcoins at $1 each only ten years ago, and never liquidated your stake. Your portfolio would be worth around $30 million today. This is the kind of potential that exists in the market for digital assets.

The market for NFTs is awash with opportunities for everyone. Whether you are a financial investor, a collector or an artist, there's something for everyone. Artists and creators, for example, can look forward to the metaverse, where NFTs will likely take center stage. Most of the tech giants—whose products have revolutionized our internet experience over the years—are heavily invested in and actively building products to align their business needs with the metaverse. We are talking about the likes of Meta Platforms (formerly Facebook), Google, and Microsoft. This is encouragement for you to be thinking along the same lines, and what better way to do it than through NFTs.

By the end of this book, you'll have learned the ins and outs of NFTs, and more importantly, understand their inherent value in our digital society.

You'll learn how artists and other creatives can now get rid of middlemen and brokers to maximize the profit they make on their hard work.

Instead of selling through intermediaries, they are now in direct contact with buyers through NFT marketplaces. This creates a win-win situation because the absence of middlemen means that buyers and creators can negotiate and agree on fair prices - even when the marketplace takes a fee themselves.

If you are not an artist or a creative, you'll find value in learning about these different NFT marketplaces and how to become a participant and trade for profit.

You will also learn about the different income opportunities available in this market, as well as investment strategies you can use to tap into those opportunities.

Perhaps one of the most important lessons you should take from this book is that NFTs represent exclusivity.

They are indivisible by design, and this is what makes them such interesting digital assets. As much as someone might still be able to create a digital copy of something like a work of art, NFTs act as proof of authenticity. Therefore, if you own an original authentic version of something, say a game, music, art, virtual property, or anything else, no one can take that away from you.

NFTs are here to stay, and mostly because they represent the value of having something that's one of a kind. They help to authenticate ownership of original digital assets, which can then be traded online using cryptocurrency.

Even if someone made a copy, they still wouldn't be able to make money off of it.

As more people get involved in NFT spaces, we can expect further growth and advancement of this concept. At the moment, most NFTs that make headlines are in the high-end market. We've seen people buy virtual plots of land on certain metaverses, expensive works of art being sold at auctions, and other exciting sales, like the sale of the first ever Tweet by Twitter co-founder Jack Dorsey.

With time, NFTs will evolve beyond these specialty or high-end applications and feature everyday activities that most people can relate to.

The age of NFTs is quite a strange time. Some of the headlines coming from the NFT market can have you rethinking the value of precious items. People are buying tweets for the kind of money you would use to buy a mansion in Hollywood. Virtual works of art are fetching ridiculous sums of money at auctions, amounts that rival the value of works by legendary artists like Vincent Van Gogh. We well and truly are in the mix of all things Crypto - let us be the first to welcome you to the Age of NFTs.

A great aspect of the NFT revolution is that it has created a platform for new artists of the modern age to become the Rembrandts of our time.

Many artists have always struggled to get their big break in the traditional art world, mostly because of gatekeepers in the industry. NFTs have changed all that. Several artists have turned their careers around by venturing into digital art, and many who were locked out of the mainstream market in the traditional art industry found an outlet through NFTs that gives them direct access to audiences.

While NFTs, and blockchain technology in general, have proven so far that there are exciting times ahead with advancement we can look forward to, we cannot turn a blind eye to the environmental impact of this technology.

Mining digital assets is a resource-intensive exercise that exerts undue pressure on our non-renewable resources of energy.

From NFTs to cryptocurrencies, the continual mechanisms implemented in different blockchain projects place a huge demand on energy consumption, with some projects notoriously consuming more than the annual energy consumption of industrial nations.

This is a conversation that we must have, given the fact that demand for blockchain assets is expected to increase in the future as we find more utility value for them in mainstream industries.

Looking back, I believe that going in on NFT art was one of the best decisions I've ever made. As an early adopter of NFT art, I've been around long enough to learn from not just other people's mistakes, but mine too. That's how I managed to create a source of passive income through crypto art.

My journey was no different from what other early adopters went through, and you might actually relate to it if you've tried your hands at NFT art before.

The short story—it wasn't easy. Around the time I started dabbling in NFT art, there wasn't as much information readily available as there is today. It was quite frustrating, especially knowing that there was a chance of losing everything while I was at it. Today, there are boundless sources of information, and you can also call on the experience of people like myself who've been in the industry long enough to offer practical advice to beginners.

When I entered the NFT industry, it was very interesting. At the time, it was all about the hype, so you can imagine how difficult it might have been to discern valuable NFT projects from hype projects that didn't have any tangible value. As I grew in the craft, I realized the need to teach beginners more about the workings of the NFT market, so you don't have to learn the hard way like I had to.

Due to the lack of simple, easily comprehensible guides, it actually took me a good two years to find my footing in the NFT market. During that time, I witnessed everything from humble beginning NFT projects to exciting hype NFTs that fizzled out incredibly fast. I've seen people lose money in NFTs, and even got close to losing some myself, save for some last-minute due diligence and research.

The good news is that it doesn't have to take you two years to figure this out. There's a lot of information available on all aspects of NFTs. Unfortunately, too much information can also be confusing.

Instead of going through all that, set aside a couple hours of your time and let this book walk you through everything you need to know about NFTs and how to make money from it. It's an emerging market that's evolving at a speed faster than most people can comprehend, so you need to stay ahead of the pack.

NFTs are a fitting contribution to a thriving global online community in which people envision a future where they can live most of their lives online. This is also heavily tied to the internet's evolution into the metaverse.

The vision is to ultimately create a metaverse where we interact online just as we would in the physical world. A world where people can work remotely, run businesses online, collaborate on projects through 3D visualization in real-time, and so on. NFTs are at the center of these advancements.

It's my hope that this book will usher you into the new age of NFTs, cryptocurrencies, metaverses and blockchain technology, and give you the necessary skills to recognize and take advantage of the amazing opportunities that will reshape the future of the internet, and ultimately, the future of humanity.

Time to deep dive specifically into NFTs. What exactly are these digital assets, and why are they so important that Collins Dictionary announced NFTs as 2021 Word of the Year?

CHAPTER 1
WELCOME TO THE WORLD OF NFTS

What exactly are NFTs? What's all the fuss about them anyway?

NFT stands for non-fungible token. They are digital assets. For something to be considered an NFT, it must be uniquely identifiable and exist digitally.

"Non fungible" means that something can't be replaced by an identical item. We will delve into the concept of fungibility later in this chapter.

The value of these assets comes from the fact that they are unique. Common examples of digital assets you might have come across already include websites, logos, and digital media content - but these aren't always necessarily unique.

As for the "token" part of NFT, it refers to the asset side of things. They are digital assets that you can buy and sell online the same way you trade in foreign currency or company stocks. One major difference between NFTs and these financial assets is that instead of a local currency, you trade NFTs using cryptocurrencies.

So, to summarize it in layman terms, an NFT is a unique digital asset that cannot be replicated.

To many people, NFTs are a new concept. In fact, many people learned of NFTs in 2020, during the Covid pandemic. However, NFTs have been around much longer than that. They've been around since 2014, but interest in them exploded at the beginning of 2021. This was around the time Twitter co-founder Jack Dorsey sold the first ever tweet for $2.9 million. Art enthusiasts will also remember this as the time Mike Winkelmann, a digital artist trading under the name Beeple, sold an NFT dubbed *Everydays: The First 5000 Days,* for $69 million. Soon after that, people began to pay more attention to developments in the NFT space.

According to a Reuters report, people have spent more than $24 billion on NFTs as of 2021. But this figure doesn't quite tell us the most important piece of information. Just a year before this, the total amount spent on NFTs was a measly $94 million. That's an astronomical increase of 25,431.9% in a year!! What's so exciting about NFTs that drove more than $20 billion in sales volume in just a year? Well, this book is here to explain it all. You will hopefully learn enough about NFTs to help you identify and seize the investment opportunities that come your way.

When you think of most digital assets that we know of, authenticity and uniqueness are flexible ideas. Someone could easily create a copy of an asset and claim it as the original version if they wanted to.

However, what NFTs offer is proof of authenticity. Let's say you created an NFT of an image and posted it online. Someone could easily download the image and save it on their computer. The difference is that while they have a copy, they cannot pass it off anywhere as the original version. You, the author, are the sole authentic owner, and there's credible proof of that, on the blockchain—another exciting concept we'll discuss alongside cryptocurrency.

Once you publish something online, anyone on the internet can view it, generally for free. So, how is it that people are spending billions on something they could easily download for free? This is where NFT authenticity makes a big difference, and will partly help us understand the astronomical amounts of money people are spending in that market.

NFTs have unique cryptographic authentication codes that can only be owned by the creator, or whoever buys exclusive rights to the NFT. This is the proof of ownership. Therefore, if someone had a copy of your NFT, it would be impossible for them to sell it as the original copy because they don't have the authentication codes.

This is a win-win for everyone. For buyers, you can take pride in buying unique content (remember non-fungible?), and no one else will have something identical to what you own. For example, you can buy a song from your favorite artist, or a special limited edition item from their collection. For sellers, dealing directly with your customers means that all you have to pay are the gas fees required in

NFT marketplaces, and not the exorbitant fees brokers and other middlemen charge on other traditional sales outlets.

Now, while NFTs and cryptocurrencies coexist most of the time, you should not confuse one for the other. Cryptocurrencies are just that—digital currency, while NFTs are digital assets that mostly prove the authenticity of something. They both, however, exist because of, and supported by, blockchain technology.

UNDERSTANDING BLOCKCHAIN TECHNOLOGY

Blockchain technology is simply a distributed, decentralized ledger on which all activities related to a digital asset are recorded. To understand how ledgers work, let's think of an accounting ledger. Accounting ledgers are the foundation of bookkeeping. They are used to record all transactions that take place in a business, and eventually, they form the basis of crucial financial statements like the balance sheet and/or the profit and loss statement for a company. *Every single* transaction is recorded in an accounting ledger. This is to help with accountability and transparency.

The same concept applies to blockchain technology, only that this is online, and the ledgers are located on servers all over the world. Every transaction that takes place regarding a digital asset is recorded on the distributed ledgers, so there's always a digital record of a transaction. If you sell an NFT, information on the new ownership is immediately created in the digital ledger.

Now, while an accountant has access to all the business ledgers and can make adjustments to them as they see fit, this is not possible with blockchain technology. In fact, everyone has access to the distributed ledger. The only difference is that unlike traditional accounting, once something is written on this ledger, it cannot be changed. This concept is called immutability, and we say the blockchain is immutable. This is how we can safely have transparency in the blockchain.

Since everyone has access to the ledger and all data on the blockchain is immutable, it's impossible to shortchange someone. This also takes us back to the earlier point we made about NFTs and authenticity. There can only be one authentic owner, and that information is coded on the blockchain. When you sell an NFT, you sell rights ownership, which is captured in the blockchain, so the buyer becomes the new authentic owner without any room for dispute.

Immutability is one aspect of the blockchain that has made it so attractive for many industries. One can instantly think of a few instances where this feature would be useful - cybersecurity and payment processing systems spring to mind.

There's no limit to the kind of assets that can be traded as NFTs. In fact, people have traded everything from copyrights and patents to music, videos, and even real estate. In short, both tangible and intangible assets can be made into NFTs. Remember that the core idea here is authenticity or proof of ownership and uniqueness.

To understand the blockchain's value, we should look no further than the construct of our digital world. Today, data is the new gold. We interact with several systems all the time, many of which are data collection points. We are surrounded by gadgets and systems that collect all kinds of data about us. This data is eventually used to model innovative and engaging solutions to real-world problems.

The business world, in particular, thrives on information. Accuracy and speed of information transfer are crucial for decision-making at the highest levels of business. This is what blockchain technology promises us: immediate access, transparency, and more importantly, shared access to information on an immutable ledger.

Certainly, even though we talk about transparency and access for everyone, you must still have authorized credentials to gain access to this information. With the proper credentials, you can see every detail about a transaction on the blockchain. If you create an erroneous entry on the blockchain, it will most likely not be validated. If it is authenticated, you still won't be able to delete or correct the erroneous entry. Instead, you'll create a separate entry to correct the first one, but both entries will be visible on the blockchain.

In essence, the blockchain creates a sense of trust, especially over limited-access networks. It also helps to fortify security through immutability, and creates a platform for enhanced efficiency.

CRYPTOCURRENCY AND DECENTRALIZATION

Cryptocurrency is digital, decentralized money that uses cryptography for security. Generally, the mention of cryptocurrency has many people thinking about Bitcoin, and for a good reason. Bitcoin happens to be the most famous and influential currency of them all. It's in the news almost every other day, and its price fluctuations elicit discussions and sometimes panic all over the world. However, it's not the only one. Other than Bitcoin, some noteworthy cryptocurrencies include Ethereum, Algorand, Litecoin, EOS, and Tezos. There are thousands of cryptocurrencies being traded today, which means that we might never even get to know all of them.

One of the key features behind the growth of cryptocurrencies is decentralization. This means that no single entity, or government for that matter, controls it. This is a unique approach to money, since we know that traditionally, the supply and distribution of money is usually controlled by central governments and banks.

Cryptocurrencies eliminate the need for middlemen and brokers like payment processors and banks, thereby facilitating transactions between users all over the world almost instantly, and at affordable rates. While some banking transactions may take hours or days to complete, it only takes a few seconds with cryptocurrencies due to the absence of systemic and institutional bottlenecks and bureaucracies.

Since cryptocurrencies are not controlled by a central organization or a government, their security comes down to decentralization and verification. Instead of central organizations, we have peer-to-peer open source networks distributed all over the world. The data is fragmented and protected through cryptographic encryption, so even with the most intelligent and powerful computers, it would be nearly impossible to attempt a hack. We also mentioned briefly the concept of immutability, which further strengthens the concept of security.

Security aside, let's look at some reasons why a decentralized currency is a brilliant idea:

ECONOMIC FREEDOM

Without interference from central banks and governments, decentralization frees a currency from national fiscal policies. You can think of financial issues like inflation and other monetary policies that could devalue a country's currency. With this in mind, we have a level playing field for everyone. It doesn't matter whether you are in the US, UK, or Mongolia, you all use the same cryptocurrency, with the same value.

It's also safe to say that decentralization protects your money from the adverse effects of inflation or deflation. The most important aspect of cryptocurrency valuation is demand and supply, and its impact remains the same all over the world.

PROTECTING USER INTERESTS

Banks can often collapse. Over the years, we have witnessed people lose their livelihoods because their favorite bank went under overnight. It gets even worse when you consider the aggressive banking policies and steep fees they charge. A decentralized currency protects users from such problems.

BORDERLESS PAYMENTS

With decentralized currencies, you don't have to worry about currency exchange rates if you need to transact with someone in a different country. This also does away with the traditional problem of limits on transactions, making international payments affordable and seamless. This also means that you don't have to worry about tariffs and exchange rates, which usually impacts import-export businesses, because eventually, the traders can pass the additional costs to the customers.

WIDER FINANCIAL INCLUSION

There are millions, if not billions of people out there who are unbanked or underbanked. These are people who, for different reasons, do not have access to traditional financial systems. What's interesting about this group is that they still interact with and move money just as much as the banked do, and in some communities, in bigger volumes.

Decentralized currencies bring such communities into the fold, allowing them to enjoy the benefits of financial inclusion. For these people, you might not need a bank anymore. All you need is a digital wallet and you can trade with people all over the world.

Ultimately, decentralization gives us an interesting picture of the power of information and technology in transforming the world, and we can only look forward to more advancements in the future.

LET'S TALK ABOUT FUNGIBILITY

Fungibility and non-fungibility are two important terms that will come up in discussions about NFTs. Learning the difference between the two should make it easier to understand the economics of NFTs.

Something is fungible when it is divisible and non-unique. An example of this is currency. If you have a dollar bill at home, and another dollar bill on your desk at work, they both have the same value. Anyone would be happy to trade dollar bills for dollar bills as the inherent value of them is exactly the same (despite them physically being different pieces of paper). The same applies to cryptocurrencies. A Bitcoin is the same everywhere. Whether you're in Texas or if you are on a safari in Tanzania, the value of your Bitcoin will not be different from someone else's.

On the other hand, we say something is non-fungible because it is unique, and more importantly, non-divisible. This means that it cannot be replaced with another, even if they look the same. For example, let's say you are attending a concert. Someone with a front-row ticket and another with a ticket at the back row both attend the same concert, and listen to the same music. However, their tickets are not the same, neither will their experiences, relative to the position of their seats.

The non-fungibility of an asset is one of the reasons why NFTs are attracting a lot of hype. Every NFT on the blockchain is unique, and this is one of the reasons why some are valued more than others. NFTs represent indivisible, unique assets, which could either be tangible or intangible. You can create an NFT of intellectual property, like a website, or programming code, or you could also create one for a video on your Instagram profile.

The most important thing is that once you create the NFT, no one can create a similar item and claim it to be identical. Blockchain technology makes it easier to authenticate the validity of your claim on the asset.

IS THE NFT SPACE A BUBBLE?

There's currently a lot of hype around NFTs - NFTs have made headlines in financial news, celebrity news, tech news, you name it. It's easy to see where this is coming from. However, we have to ask ourselves whether this hype belongs to a bubble that is about to burst, or if NFTs will stand the test of time.

Well, the first step in figuring out whether something is just hype or if it has substance is to understand its inherent value. What's in it for people who buy NFTs?

Fundamentally, an NFT has no value, other than the fact that it represents the authenticity of ownership. This explains why you can still download an online photo of someone's NFT and share it within your wider networks, and the owner won't even care about it. So, in short, we can still say that value lies within the item for which the NFT was created, and not in the NFT itself. The NFT simply gives us a means of authority on the blockchain. If you sell a house NFT, for example, your house won't be coded into the blockchain. The NFT that represents your house, however, will be.

Whether NFTs are a bubble or not was always a discussion that we were going to have at some point. But, before we jump into it, let's take a lesson from how some similar discussions have turned out. The best example is the internet. Some people once thought it was a farce. They thought it would never replace the physical way of doing things, yet today we practically live most of our lives online.

Given the proliferation of NFTs into different spaces—it's not just limited to celebrity art anymore—we can expect more uptake of NFTs, and even more discussions around the long-term viability of these digital assets. Unfortunately, as more people buy into the idea, there's also the risk that quite a number will be scammed.

Proof of this is in the Bitcoin and cryptocurrency market in general. So far, people have won and lost in equal measure. Some have built fortunes while others have lost almost everything they invested. In other words, the risk in NFTs does not belong to the creators, but to the holders. What are the odds that you'll wake up one morning and are left holding a worthless image?

Bubbles are more common in our society than you would initially think. It generally means there's a ton of interest in something. New technologies have been at the forefront of modern bubbles, but at some point, the market always finds a way to correct itself and restore normalcy. With that in mind, however, there are valid reasons to believe that NFTs might actually beat the bubble hype.

If we look at the basic concept of NFTs, they tackle important issues that have plagued our society for years—brokers and middlemen. Without people taking their cut for connections, many transactions will be considerably cheaper. There's even talk of NFTs helping in the fight against online piracy. So, in essence, for straightforward and honest people, NFTs are a welcome advancement. It also brings peace of mind to know that someone can't sell something you own and claim they are the legitimate owners.

There's also the fact that NFTs have a significant role to play in the biggest development on the internet at the moment, the metaverse, which is fronted as the next stage in the evolution of the internet. The usual suspects—Meta Platforms (formerly Facebook), Google, NVIDIA, Intel, Microsoft, and so on—are the notable brands creating the metaverse, or their individual versions of it. On the face of it, you can buy into their idea, since these brands are constantly forging new pathways for society with their technological advancements. Having said that, one could argue that Google Glass was a catastrophe. So perhaps this tells us to be more cautious with NFTs and make slower, calculated moves.

From a broader perspective, it seems NFTs will hold the fort and stand the test of time. This is because their appeal has transcended the traditional niche concept, and even non-techy people or digital enthusiasts are deriving value from NFTs. In particular, we have a young generation that is getting exposed to NFTs, and will grow with this concept as a normal thing.

Widespread adoption of NFTs in the visualization and gaming environments is another reason why we might see NFTs going the whole way. NFTs have actually drawn attention to cryptocurrency technologies from people who previously had no interest or relationship with them. To interact with NFTs in whichever capacity, you need a digital wallet. Just like that, your entry into crypto spaces is guaranteed, and so begins your interest in unraveling the marvels that be; yet consciously, you may never have been interested in cryptocurrencies.

At the moment, it is indeed difficult to unlink NFTs from hype, but that also depends on the kind of user you are, or the nature of your interest in NFTs. There's so much innovation around NFTs that it's almost impossible to imagine that this space could crash and burn. Yet, stranger things have happened in the digital world before.

CHAPTER 2
A BRIEF HISTORY OF NFTS

NFTs have grabbed the headlines, with people spending millions of dollars to own them. There are many whose transactions never made headlines, but they still fetched large sums of money. The course of this chapter is to help us track the origins of NFTs from its humble beginnings, to some of the most expensive NFTs we've come across today. Also note that as the NFT space is still evolving, there's plenty more yet to come, especially as we usher in the age of the metaverse, where the roles and functions of NFTs will be vitally important.

THE COLORED COINS

As far as the history of NFTs is concerned, the starting point remains a speculative issue. Some people believe that the earliest NFTs came to be around 2017, the year in which the NFT market started attracting a lot of attention. However, the earliest digital assets that we can classify as such, given the definitive features of NFTs that we know today, are the Colored Coins that were created in 2012-2013.

The Colored Coins were small Bitcoin denominations that could be passed off as a Satoshi (the smallest Bitcoin unit). Like modern NFTs, Colored Coins were created to represent value in digital assets. They could, therefore, be used to represent value in tokens, ownership of company stocks, coupons, and property. You could also create your own cryptocurrency and use the Colored Coins as representative assets.

As the Bitcoin blockchain was already making headway around this time, the Colored Coins helped to advance the functionalities of the Bitcoin blockchain. Unfortunately, the only way Colored Coins could be used to represent value in digital assets was if all parties involved agreed on their value.

Without full consensus, the Colored Coins valuation system was unworkable. For example, let's say five people involved in the sale of a house agreed that 5,000 Colored Coins were equivalent to the price of a beachfront property, and the sale was to be completed on these terms. If any one of the parties changed their mind, the entire transaction could not take place.

This challenge highlights why *smart contracts,* as introduced in the Ethereum blockchain, were a brilliant idea. Smart contracts are self-executing contracts. The terms of the contract are coded into their structure on the blockchain, and the contract is executed as soon as the preconditions are met. Therefore, assuming that the five parties who intended to sell the house had a smart contract, all that needed to happen was for the terms of the contract to be met. Even if one of the parties changed their mind as to the value of the Colored Coins, as long as their change of mind wasn't a precondition for honoring the smart contract, the sale agreement would have been honored.

Another problem with Colored Coins is that they could only thrive in an ecosystem where permissions had to be sought and granted. This goes against the concept of a trustless network which is one of the foundations of blockchain technology. Considering the potential chal-lenges involved in a permission-based system, Colored Coins effectively failed in their approach to represent real assets in the blockchain. The good news, however, is that they laid the foundation for further research and experi-ments that eventually yielded the NFTs we have today.

COUNTERPARTY

Counterparty picked up on the lessons and flaws of Colored Coins in 2014. At this time, the value of blockchain was becoming apparent, especially in terms of assigning real physical assets representation on the blockchain. At this point, Bitcoin was the only notable blockchain that most people could consider investing their time and money into. However, the Bitcoin blockchain wasn't capable of handling anything other than digital currency operations, and this is how Counterparty came to exist.

The solution: an open-source, distributed financial system that built on the Bitcoin blockchain to harness the power of its peer-to-peer functionality. It was built as a wholesale decentralized exchange, complete with digital assets.

In 2015, gaming enthusiasts were treated to a glimpse of merging blockchain technology in the modern gaming environment. "Spells of Genesis" leveraged Counterparty and began offering in-game assets built on the blockchain. Not only did they pioneer in-game assets on the blockchain, but they also were the first company to raise an initial coin offering (ICO). An ICO follows a similar approach as an initial public offering (IPO), only that it involves users crowdfunding to purchase blockchain assets. "Spells of Genesis" also created a native token to be used as currency in the in-game ecosystem, called BitCrystals.

A year later, "Force of Will", one of the most popular trading card games, used Counterparty to release their cards. This was a critical moment in the evolution of gaming and blockchain technology in that this was the first time a big gaming company, without any history in cryptocurrency or blockchain in general, put their weight behind blockchain technology to support gaming assets. The entry of "Force of Will" into the blockchain world brought a lot of attention to the gaming world, and users became more receptive to the idea.

The next stage in the evolution of NFTs is one that was always going to be a matter of time: memes. Memes have come a long way. They are as big a deal in the gaming world as they are everywhere else they've been introduced. Many people find the witty quotes and punchlines in the memes relatable. Mentioning memes and the blockchain today will almost automatically have you thinking about Dogecoin. However, Rare Pepes were the pioneers of the NFT meme space in 2016. The meme, featuring a frog, is so popular it has a directory supported by a loyal fan base.

GOING MAINSTREAM

By 2017, Ethereum's popularity was rising in blockchain spaces. People realized that they could overcome most of the limitations of the Bitcoin blockchain by creating their projects on Ethereum, whose functionalities support many developer projects. This is how memes made the crossover into the Ethereum blockchain. A decentralized marketplace for memes and trading card games, Peperium, was born. The selling point for this marketplace was that users were able to curate memes that would live forever on the Ethereum blockchain. Rare Pepes became an instant hit on Ethereum.

Soon after the success of Rare Pepes on Ethereum, Cryptopunks made their entry into the digital space. The origin of these NFTs was Cypherpunks, one of the early iterations of digital currency in the 1990s, from which Bitcoin's roots can be traced. When the Cryptopunks project was launched, users were allowed to claim one of the 10,000 creations for free. Soon after, these became popular in secondary marketplaces where they were traded.

CryptoKitties was released in 2017 at the ETH Waterloo Hackathon, one of the biggest hackathons in the world for advancing the Ethereum project. In the presence of more than 400 developers, this was the perfect audience to release the project, whose dev team ultimately won first place. CryptoKitties became an instant hit, and went viral.

Perhaps a case of the perfect blend of timing and opportunity, the viral spread of CryptoKitties happened around the time when the crypto market was bullish. There was so much excitement about blockchain projects at this time, which further fueled the rise of not just CryptoKitties, but many other blockchain projects that existed at that time. Buying, breeding, and selling digital cats was the craze, and from there, the value and potential of NFTs became apparent.

Between 2018 and 2019, there was a flurry of activities in the NFT environment. Hundreds of exciting projects were created, and many others were in development. NFT marketplaces like SuperRare and OpenSea flourished, and have maintained the growth trajectory to date.

Around this time, many gaming franchises joined the NFT ecosystem, further spiraling growth and excitement about this project. Given the degree of advancement in the gaming environments, the metaverse discussion soon picked up, further highlighting the role and value of NFTs.

Celebrities, artists, and musicians also realized the value of NFTs around this time and soon started tokenizing anything possible, from live shows to artwork and music. The creative economy came face-to-face with a massive disruptive force in NFTs.

In 2019, Nike patented CryptoKicks. The patent allowed Nike to create digital copies of their footwear on the blockchain. Like CryptoKitties, CryptoKicks are coded in a manner such that they can be bred and the offspring shoes mutated into unique virtual shoe projects. That's not all; the resulting variants could even make it to production, and end up in stores as physical shoes. Nike also took advantage of the gamification of NFTs, allowing the use of CryptoKicks in video games.

Fast forward to 2021, arguably the most exciting year for NFTs so far. This was the year digital artist Beeple's work was auctioned for a staggering $69 million, Twitter founder Jack Dorsey's first tweet sold for $2.9 million, and many other NFT sales that brought a lot of attention to this ecosystem.

The biggest distinction between NFTs at this point and the earlier iterations, was that while the early stages were mostly driven by hype around the projects and the Ethereum blockchain, the NFT marketplace was now becoming flooded with assets of tangible value. The dynamic of NFTs was well and truly changing. You could now actually buy real estate NFTs. This also coincided with a period where most of the big tech companies were actively investing in building their version of the meta-verse, commonly agreed as the next big leap in the evolution of the internet.

Even though there are many cool projects in NFT market-places at the moment, you get that feeling that the best—especially in terms of the true utility value of NFTs—is yet to come. It's even more exciting knowing that you'll be right there, fully in the know, when it happens.

MOST EXPENSIVE NFTS

2021 was an interesting year for blockchain assets. The tremendous growth in this sector sparked a lot of attention from several interested parties, both in developing the infrastructure on which blockchain assets thrived, and in availing the necessary capital to push blockchain ventures to the next level.

Notably, NFTs were among the digital assets that got the most attention on the blockchain. Conversations about blockchain technology were everywhere, and with NFTs making headlines for raising crazy amounts at auctions, suddenly, everyone wanted in on the action. It went from a marketplace that people barely knew anything about to one where billions of dollars were exchanged for digital assets. Let's have a look at some of the most expensive NFTs that have been in the headlines so far:

BEEPLE'S CREATIONS

Mike Winkelmann, the digital artist behind the pseudonym Beeple, is one of the content creators whose work has gained a lot of attention in NFT spaces. *Crossroad*, which sold for $6.6 million, is an artistic display of former US President Donald Trump lying down on a field while people around him go on with their lives, oblivious of his presence.

Another piece of Beeple's work fetched $29.98 million. *Human One*, the dynamic digital sculpture of a wandering astronaut in different environments, is an interesting NFT because according to Beeple, his intention was to ensure that the NFT was continuously updated throughout his life-time, ensuring it would never be a static NFT.

Everydays: The First 5000 Days is another of Beeple's famous works that eventually sold for $69 million. The amazing thing about this sale (apart from the price tag, obviously) is that the starting bid was a meager $100! Since Beeple had already become a household name in NFT marketplaces, it was only a matter of time before the larger bids started streaming in.

From an opening bid of $100, people were bidding at more than $1 million within an hour, until the NFT was finally sold for $69 million in February 2021. *Everydays: The First 5000 Days* will always remain unique in that it was the NFT that got mainstream audiences interested in the NFT marketplace, and soon after, many people started exploring the possibility of creating and selling their own NFTs.

NOORA HEALTH – SAVE THOUSANDS OF LIVES

Noora Health is a care giving organization that improves the lives of people at risk in South Asia. They launched an NFT project that fetched over $4.5 million when it sold for 1337 ETH. The proceeds from the sale were committed to saving the lives of newborns.

THIS CHANGED EVERYTHING

Sir Tim Berners-Lee (creator of the World Wide Web) auctioned the original source code for what we now know as the internet as an NFT. The aptly-named 'This Changed Everything' NFT was auctioned for $5.4 million in 2021 to an anonymous buyer.

ART BLOCKS COLLECTION – RINGERS #109

The Art Blocks platform is specifically for immutable on-demand content on the Ethereum blockchain. The concept is simple: if you find something you like, you pay for it and receive a digital version of it in your Ethereum account. There's a variety of content you can create using this platform, including 3D models, images, or interactive experiences. Ringers #109 is a collection of 99,000 Art Block NFTs that sold for 2,100 ETH, fetching $6.93 million in the process.

RIGHT-CLICK AND SAVE AS GUY - XCOPY

This meme NFT fetched $7 million, the equivalent of 1,600 ETH at the time of sale. The meme was purchased by legendary hip-hop artist Snoop Dogg. This NFT was created as a mockery of people who don't believe in the authenticity and value of cryptographic art, or any form of internet art for that matter. Since anyone can easily download the meme, it's become a popular sarcastic meme used to call out those who don't believe there's value on the blockchain.

CRYPTOPUNK – LARVA LABS

CryptoPunks made their mark in history as one of the earliest NFTs ever created. Larva Labs created 10,000 random portraits so unique that none of the generated characters look alike. Each of the random portraits were eventually claimed on the Ethereum blockchain.

CryptoPunks were originally sold for 2 ETH each, which at the time was around $1,200 each. At the moment, users can trade them in secondary markets like OpenSea, where they can be exchanged for Ether, Bitcoins, or any other cryptocurrency.

These NFTs became quite popular in the market when they were first released, with the earliest bid being $2 million, and eventually selling for around $7.5 million. Some of the more famous CryptoPunks traded so far were #3100 and #7804, both belonging to the Alien Punks series, a collection of nine rare aliens. Of all the CryptoPunks created, #3100 is one of only 406 that had headbands. Later in the year, 2021, #7523, popularly known as the Covid Alien, sold for $11.75 million at an auction.

#7523 held the title as the most expensive CryptoPunk ever sold, until February 2022 when #5822 was sold for close to $24 million. Prior to this, there was the curious case of #9998 that apparently sold for 124,457 ETH, or approximately $530 million, but turned out to be a publicity stunt, as the owner had simply bought the NFT from himself through an elaborate scheme that involved obtaining a loan through smart contracts, and transferring the NFT to one of his wallets.

PAK – THE MERGE

This is a unique piece of artwork made up of fragmented art that was sold between 2nd and 4th December 2021 for a sum total of close to $98 million. What's unique about this artwork is that instead of being purchased by a single owner, 28,983 collectors owned the NFT. *The Merge* was sold in fragments known as mass, so by the end of the auction, more than 266,000 masses of *The Merge* had been auctioned. Bidding for each of the fragments opened at $575, with the token price increasing by $25 after every six hours until the final sale was completed.

MOST FAMOUS NFTS

Many NFTs have been traded online since NFT market-places went mainstream in 2021. We highlighted some of the most expensive NFTs ever sold in the previous section. However, even though these fetched mouth-watering sums, not all of them can be considered famous NFTs. There have been some interesting NFT sales that made headlines, became viral stories, and might remain NFT folklore for years. Let's briefly discuss four such NFTs, and through their significance, understand the reasons behind their newly acquired fame.

NBA TOPSHOT

Sporting activities generally have some of the most loyal fan bases in the world. Their loyalty is so fierce that there's no limit to what sports fans can do for their teams, or to support a game they love. A good example is the "NBA TopShot" NFTs.

NBA TopShot is the brainchild of Dapper Labs and the NBA. TopShot is one of the early NFT projects that helped to bring mainstream audiences to NFT marketplaces. Through TopShot, Dapper Labs created an outlet for basketball fans to trade NFTs of basketball history, which in essence meant they were actually writing an important chapter in the history of NFTs. TopShot has users sharing and trading everything about the NBA from their favorite moments, special occasions, players, and favorite teams. Anything of sentimental value to NBA fans is worth something to someone out there, and that's how TopShot managed to raise awareness of NFTs.

TopShot is similar to trading game cards, only it features a combination of digital art and NBA highlights. Dapper Labs are expressly licensed by the NBA to create and sell footage of NBA highlights, which are then attached to NFTs. Dubbed 'Moments', these special clips are actually what makes TopShot so appealing to NBA fans. The NFT depicts the rarity of the moment in the highlights, which come with a unique digital serial number as proof of authenticity on the blockchain.

Following the concept of digital scarcity, Dapper Labs only produces a limited number of Moments NFTs, which makes them more valuable to collectors. Naturally, the most expensive Moments are those considered the rarest by the fans. The ultimate goal of the NBA TopShot project is to create a new gaming experience for fans beyond trading in special occasions, collectibles and moments. Users can compete and win prizes on the blockchain.

There are five different categories of Moments on NBA TopShot:

- **Common Moments**

These make up the bulk of NBA TopShot Moments. They are widely available, and among the easiest Moments NFTs users can access. They make up more than 96% of the Moments NFTs ever created. It's common to find tens of thousands of a single version of an NFT in the marketplace.

- **Fandom Moments**

These are special highlights for special game events. They usually come with special instructions. For example, one of these NFTs might be exclusively available for sale to fans who attended a certain game. They generally correlate with special occasions.

- **Rare Moments**

These make up around 2% of the total number of Moments NFTs created, so no more than 5,000 NFTs. They are exclusive NFTs, and considered among the most difficult NFTs to obtain. Rare Moments are mostly reserved for historic replays featuring NBA legends.

- **Legendary Moments**

These are the most extremely rare NFTs in this space, as they make up around 0.1% of the total Moments NFTs created. That's around 50-500 Moments NFTs.

- **Ultimate Rarity Moments**

This is a special category of NFTs that is only available at auctions as a limited edition featuring one or three NBA moments.

Owning one of these Moments gives you the right to use it as you would any investment on the blockchain. For example, you could sell it on the secondary market or simply simply hold onto it for sentimental value.

CRYPTOKITTIES

Like NBA TopShot, CryptoKitties is significant in the history of NFTs. It was, after all, the first instance that the Ethereum blockchain was used for mainstream activities. CryptoKitties uses Ethereum's smart contract functionality to allow users to exchange kitties on the blockchain securely.

The CryptoKitties concept is simple—users breed and sell digital cats on the blockchain as collectibles. With more than a million active users, CryptoKitties has witnessed millions of dollars in sales over the years, thanks to the fact that it allows users to combine gaming and making money.

The first kitty ever sold was Dragon, for 600 ETH, and is considered one of the most expensive kitties ever sold on the platform. Other valuable kitties on the platform are the Founder Cats, also known as Generation 0 Cats. They are quite rare and therefore command a hefty price tag. The success of CryptoKitties gave rise to other collectibles on the blockchain, like tiny monsters, pandas, and puppies.

JACK DORSEY'S FIRST TWEET

For most people, a tweet is just that, a 140-character message that expresses something. It might be a meme, a promotional message, a chat with your friends, or even part of a thread of tweets telling a story. Twitter founder Jack Dorsey's first ever tweet, however, was more than that. It was worth $2.9 million at an NFT auction.

The five-word tweet that read *"just setting up my twttr"*, was auctioned for charity, with the winner likening it to buying the Mona Lisa. Twitter has made significant strides in our lives over the years, being the go-to platform for everything from activism to product marketing and promotional events. You can actually follow trending stories and watch news unfold on your Twitter timeline instead of necessarily having to wait to watch prime time news. Given Twitter's prominence in our social lives, it's easy to see why someone would liken the first ever tweet to a digital age Mona Lisa. As the NFT was sold in Ether, Jack's intention was to convert the amount to Bitcoin and donate the proceeds to the Africa Response Fund by Give Directly.

Following this sale, blockchain enthusiasts believed that this would open up opportunities for other people to sell online posts, especially influencers, musicians, and celebrities. Due to their huge following online, it's possible that ardent fans might feel owning something like a tweet gets them closer to their favorite celebrities. The concept is similar to the sentimentality of buying collectibles and physical memorabilia.

BEEPLE

Beeple is one of the most well-known digital artists of our time, and that is despite the man not even considering himself an artist a few years ago. His exploits in NFT marketplaces have encouraged many artists to keep pursuing their dreams, and to not give up trying to monetize their craft.

For the 13 years leading up to his fame in the NFT marketplaces, Beeple completed a drawing every single day, mostly with pen and paper. His project, *Everydays: The First 5000 Days*, was the culmination of 13 years of work that made headlines when it was auctioned at $69 million. Beeple's newer projects are now mostly digital in nature.

According to Beeple, NFTs have sparked a revolution in the art industry. Prior to his fame, he felt the world of art was quite dismissive, and a difficult place for artists to earn an income. However, blockchain projects like NFTs have reversed this, giving artists an upper hand by eliminating the traditional gatekeepers in the industry and providing them the means to sell their art directly to customers.

One of the challenges that has plagued the art industry for years is the aspect of flipping. Just like flipping homes—

where you can buy a house, make some improvements and sell it at a profit—collectors do the same, except they don't even make any changes to the art. All they do is buy the work from an artist and resell it.

The blockchain approach to this trend—through NFTs—gives artists more power and control over their work. Once sold, the NFT includes a license that stipulates the terms of use in the future. For example, users buying Beeple's work are made aware that he will earn 10% off the resale of any of his works in secondary markets. Instead of collectors enjoying all the benefits, this method helps artists stay connected to, and earn from their work in perpetuity.

It's not just the artist who stands to benefit. For example, another digital artist, Sara Ludy, includes a 7% profit clause in her work, which is distributed among the employees in her gallery. While this act is laudable, we must recognize the fact that this is the trickle-down effect of Beeple's success in the digital art space. Artists finally have an upper hand, and a say in what happens to their work in the market.

CHAPTER 3
HOW THEY WORK - LOOKING UNDER THE BONNET OF NFTS

"Never invest in a business you cannot understand. You have to learn how to value businesses and know the ones that are within your circle of competence, and the ones that are outside".

WARREN BUFFETT

These wise words of arguably one of the finest investment-minded individuals in modern history sheds a lot of light on the value of this book. Let's face it, NFTs, cryptocurrency, and other blockchain projects are the investments of our time. They are happening right now, and will continue to reshape the future of financial investments. In order to keep up with the changing dynamics of the financial markets, you must learn.

Knowledge of how NFTs and the blockchain system work will make it easier to understand their value proposition. To figure this out, we have to go all the way back to the origins of money.

The origins of money can be traced back to our ancestors bartering for different goods. The legitimacy of this claim has been called into question from time to time, but whether the barter system did exist or not, we learned about how it worked, and from there, we can draw some parallels to explain the value of NFTs.

Exchanging one item for another, say a sack of wheat for a bag of rice, had a lot of loopholes. One of these was the difficulty in tracking what others owed you, or what you owed them. Ideally you would trade at exactly the same time, but if not possible, you'd have to keep a mental note of these tiny details about barter transactions, which could be cumbersome and inaccurate.

Over time, the need for a recognizable means of payment became apparent, and that's around the time people started accepting things like salt and beads as forms of payment. These were acceptable for some time as they provided the earliest form of what we could consider a financial ledger system. How these evolved into the currency we have today comes down to the unique qualities of money, and how each of these qualities met the needs of users at the time.

For something to be considered or accepted as money, it must meet the following conditions:

- **Scarcity**

Scarcity and value go hand in hand. This concept has been widely used to explain the inflationary values of many cryptocurrencies. But it also applied to the earliest forms of money.

For example, assuming that we could use leaves on trees as money, anyone could climb a tree, pluck some leaves and pay their debts, or buy whatever they wanted. This simply erodes the value of the currency and makes it worthless. Like our fathers always told us though, money does not grow on trees.

As money evolved through different forms, you realize that even commodities like salt and beads were scarce during the periods in question. Further down the evolutionary line, we see the introduction of precious metals like gold, whose very existence was a rarity, making them even more precious to both those who had them and those who didn't but needed them.

- **Durability**

Anything acceptable as money should be durable. A good example is the 'paper' currencies we use today. Countries like Australia have all their bills printed in plastic so that they are more durable. The UK is starting to print their newer bills in this plastic material for longevity too. It is quite remarkable that even paper notes of the past survived for so long given they were so susceptible to wear and tear.

Nowadays, A single plastic note can exchange hands multiple times in a day without being defaced. We see this concept in the precious commodities that were widely exchanged as money, and also in cryptocurrency. There's no chance of decay or corrosion.

- **Divisibility**

Money must also be divisible. Just as we can break down a $100 bill to quarters and pennies, we can also do the same for cryptocurrencies. Bitcoin's smallest unit, for example, is the Satoshi. For the precious commodities once accepted as money, you could distribute them in smaller units. For example, instead of exchanging a bucket of precious beads for a loaf of bread, you could instead offer a few beads in exchange for just the flour. The point here is the ease of divisibility.

- **Cognizability**

Money must be recognizable in any form. If you visit a foreign country, you might not know or understand their currency, but you'll recognize it on sight and know that it is money. Perhaps someone might explain its value to you, and the ratios of divisibility, but you'll never mistake it for something other than money.

This universal aspect of cognizability is evident in cryptocurrencies. The world over, it is commonly known what Bitcoin is about, and that is it different from Ether or any other digital currency. Ultimately, the universally accepted forms of money are always cognizable, with unique features that can be verified for authenticity.

- **Fungibility**

Fungibility represents the ability to exchange one item for another, of the same type and characteristics. If an item is fungible, it means there's no difference in value with the other item for which it can be exchanged. This is how fungible assets support trading activities.

Even though it is a definitive characteristic of money, money is the simplest example of fungibility. You can distribute a dollar bill into four quarters, and as long as you have each of the quarters, their value will always be a dollar—you can even use them in place of a dollar to pay for something that you'd normally pay a dollar for. In this case, we've split the dollar into quarters, yet its fiscal value was unchanged.

Fungibility also reiterates the fact that all money is the same. Whether you have $10 or two $5 bills, you still have $10 in total. Let's say your friend loans you some money, maybe $100. It would be absurd to expect that when you pay them back, you do so in the exact bills they gave you. All they need from you is a total of $100 back. You can send a wire transfer, pay back in cash, send it through online payment systems, or any other form, as long as you pay back $100.

- **Transportability**

Money should also be easy to carry around. This was one of the drawbacks of the barter system. It was cumbersome to carry whatever you needed to exchange everywhere, hoping that you'd also meet someone who was willing to exchange with you in kind, and in the quantities you desired.

Modern currencies fit this profile. You can walk around with bills in your pocket, and no one would notice. You can transfer millions from your bank account through an app, without having to walk out of the bank with suitcases full of cash. We also have third-party payment systems like PayPal that have streamlined ecommerce such that you can instantly transfer money to a business partner in a different country.

Digital currencies also meet and improve on this threshold. For example, you can save Bitcoins on your phone as a text message. Anyone who comes across it might not understand what the random alphanumeric text message is, but you know it's your money.

As far as money is concerned, the introduction of gold was a game-changer. Not only does gold meet all the qualities of money above, but it has also acted as a store of value for centuries. This also explains why the currencies we've used since the introduction of modern money were once all backed by gold reserves.

Cryptocurrency, and Bitcoin in particular, was built on these features, and more. For example, Bitcoin takes the concept of scarcity to a new level. There will only ever be 21 million Bitcoins in circulation. This is another reason that makes it a scarce commodity, as users scramble to try and mine the next block on the blockchain. When the 21 millionth Bitcoin is mined some time in the 22nd century, mining fees might no longer apply and instead, users could only earn from transaction fees. Creating such a fixed supply of Bitcoins was also a smart move aimed at preventing inflation.

THE CASE FOR CENTRALIZATION

While decentralization has its benefits, especially in the blockchain world, centralized systems aren't necessarily as bad as they've been made to look since blockchain conversations went mainstream. On money matters, centralization has its benefits, which makes you appreciate the status quo, even if you can see its inherent flaws.

One of the reasons why financial systems have served us well over the years is because of the efficiency of centralized systems. The current money system has served generations smoothly, because at times, it's good to know that there is stability at the very top, especially when your money is concerned.

Central banking institutions have held the global financial economy intact because of a clear hierarchy and established order in the financial system. Banks and other financial institutions play their roles within ethical and professional boundaries set by the industry watchdogs. Without central oversight, the financial sector could be (more of!) a mess.

Centralization also makes it easier to run and manage systems. For example, your scheduled system maintenance and other planned issues in advance. Banks generally inform customers when their systems will be down for maintenance, and provide alternatives that customers can use in the meantime. It's the same thing with hardware deployment. It's easier to plan for what you know, helping you lower costs in the process, than the alternative of decentralization, where no one knows anyone, and we must trust that everyone is acting in good faith, for the greater good of the entire financial ecosystem.

So, when you think about it, decentralization might be ushering in a new dimension in how we handle money, but centralization still keeps some sanity in the global financial ecosystem, despite its unique challenges.

HOW THE TECHNOLOGY WORKS

We just highlighted a strong case for centralization, and why the status quo has remained so for years. While centralization has mostly worked for the banking industry, it has some unique challenges that we shouldn't really be struggling with in the 21st century. The banking system, for example, highlights some of these challenges brilliantly.

There are issues that almost every one of us has experienced with their bank from time to time. A technical hitch that makes it difficult to complete your transactions; perhaps someone sent you some money, but it's yet to reflect in your account because of a technical delay—or maybe you sent the money to the wrong account, and now the bank is taking forever to trace that money and route it to the right account.

Maybe your account was hacked, or you were unable to complete a transfer because the bank imposed transaction limits on your account. Perhaps one of your friends in another country wants to send you some money, but your bank won't allow the transaction unless you provide further information about the international bank, and fill in more paperwork, taking more time out of your busy day. They might even flag the transaction as suspicious.

Unfortunately, these experiences are as common in the modern banking system as they are frustrating, and this is why we need to understand how the cryptocurrency system works, and the value we can derive from it.

Most of the challenges highlighted above can be avoided by using cryptocurrency. The fact that there's no central authority governing your transactions already eliminates the procedural delays common in the banking sector. Furthermore, the automation of transactions on the blockchain makes them much faster than with a traditional banking system.

All cryptocurrency transactions are written immutably on the blockchain. The permanence of these transactions makes it impossible for anyone to alter the records, so counterfeiting is one problem you won't have to deal with. Your friend or relative in another country could simply send some money to your digital wallet, which would be confirmed in minutes, without worrying about currency conversion, or the transaction not going through because of transaction limits imposed on your account.

Every crypto transaction involves two important keys: the public and private keys. The public key is your public address that is visible to everyone. This is what people use to send you money on the blockchain. Your private key is like your password. You use it for decryption, and without it, you won't be able to access your account.

HASHING

When someone sends you money, your private key digitally signs the transaction, confirming its legitimacy on the blockchain. Each cryptocurrency project uses a unique hash function, like the SHA-256 algorithm for Bitcoin, and the Ethash algorithm for Ethereum. Hash functions are used in hashing, a process of data transformation through a complex mathematical formula. Hash functions help to make blockchain transactions safer and more secure. Bitcoin, for example, uses the SHA-256 algorithm twice, a process known as double SHA-256. Hash functions are one of the reasons why it's extremely difficult for someone to hack the blockchain.

PRACTICAL BLOCKCHAIN APPLICATIONS

Blockchain applications have often been criticized for not having practical value. This argument often elicits mixed reactions from different factions, each with valid arguments for or against it. In particular, there's the argument that people spend so much money to advance technology that is, at best, contributing to the continued degradation of our environment. Several reports have often indicated that Ethereum and Bitcoin alone consume more power than the combined consumption of some industrial nations. Yet we still spend so much money on advancing the distributed ledger technology. For context, here's a list of the top cryptocurrencies by market capitalization as at 1st May, 2022 according to Forbes:

1. Bitcoin (BTC) $723 Billion
2. Ethereum (ETH) $333 Billion
3. Tether (USDT) $83 Billion
4. Binance Coin (BNB) $62 Billion
5. U.S. Dollar Coin (USDC) $49 Billion
6. Solana (SOL) $29 Billion
7. XRP (XRP) $29 Billion
8. Terra (LUNA) $28 Billion
9. Cardano (ADA) $26 Billion
10. TerraUSD (UST) $19 Billion

Note that there are thousands of cryptocurrencies currently being traded in a market whose overall combined value is worth more than $2 Trillion.

There's got to be more to a tech industry worth trillions of dollars than just acting as an alternative form of currency. This is where the true utility value of blockchain technology exists. Beyond the well-known financial concept, the use of blockchain technology has been welcomed in several industries. There's the case of Walmart, who solved their high refunds and returns problem by implementing blockchain solutions in their supply chain.

Given their elaborate supply chain, it wasn't easy to identify the exact point of failure resulting in the high returns. The problem could have been anywhere from production, transportation, storage or even in their retail sector. A blockchain solution was a viable option, given the immutability. Data would be entered into the blockchain at every step in the supply chain. Since the data is permanently written on the blockchain, it became easier to identify the sectors where problems arose, and implementing real-time solutions eventually helped them tackle the problem of high returns.

Apart from the retail sector, blockchain immutability has also been implemented in voting systems. For example, Estonia implemented a digital voting system, using encryption protocols to assure voters of security and authenticity throughout the process. With blockchain technology, they take verification a notch higher, ensuring that only those who are legally authorized to vote get to do so.

We also have the Abu Dhabi Securities Exchange (ADX) that approved a voting system built to leverage blockchain functionalities in their stock exchange. Using this system, investors and other stakeholders can easily follow proceedings at annual general meetings (AGMs). This was one of the UAE government's initiatives to modernize their economy and attract more investors. This came around the time when blockchain utility in the Internet of Things (IoT) and other forms of governance was already being tested in Dubai.

Finally, blockchain utility might come in handy for inter-company or intra-company transactions. Entities that do a lot of business together generally spend a great deal of money on banking charges to process transactions. This can be avoided by setting up a blockchain-based payment system. Other than avoiding huge transaction costs, this would also make it easier for these companies to reduce the amount of time spent processing the transactions, and unnecessary paperwork.

BENEFITS OF NFTS

As there are different types and categories of NFTs in the marketplace, it's important to understand their benefits before you commit your money.

- **Ownership**

First, there's the issue of ownership. Ownership history carries some clout, and there's value in that. For example, and as mentioned previously, Beeple is one of the most famous creators in NFT spaces, both for his role in the history of NFT sales, and also because he's an amazing artist. So, anything that comes with his name attached is valuable.

Still on the subject of ownership, NFTs are also valuable because of validation of ownership. As we mentioned earlier, anyone could download an image of an NFT. However, the difference is that none of those who have the downloaded copy own the NFT. What they have is a worthless image on their devices. The authentic owner, however, has their NFT validated on the blockchain, and is the only one who can profit from the NFT by selling it.

- **Future Value**

Even though NFTs are popular right now, there's arguably more potential in their future value. Owning an NFT right now sets you up for a potential windfall in the future when their full value is fulfilled. There's a lot of talk about the metaverse at the moment, and most of the big tech companies are involved. When the metaverse fully comes to fruition, NFTs are poised to play a significant role in the commercial environments within the metaverse.

- **Utility**

Utility is about what we can do with the NFTs. There's so much utility value in NFTs at the moment, and yet we have only scratched the surface. Some NFTs provide membership access to exclusive activities like concerts. Gamers, for example, can use some NFTs to purchase in-game assets across different gaming franchises. NFTs are also used in decentralized autonomous organizations (DAOs), which are basically a group of users pooling funds to pursue a common objective.

- **Scarcity**

Ultimately, many users are investing in NFTs because of their scarcity. Almost every popular NFT environment has some rare NFTs that are considered more valuable than the rest. We discussed the concept of scarcity and its role in perceived value of blockchain assets earlier in the book, so you already understand how it affects the NFT market. We can, however, delve into the psychological aspect of it, to learn more about why, and how we value things as we do, in the next section.

HUMAN PSYCHOLOGY AND THE PERCEPTION OF VALUE

The issue of scarcity and blockchain assets gives us a glimpse into how the human mind works. Practically speaking, NFTs are unique digital assets in that they are things that we normally wouldn't assign value to. Take the case of famous NFTs that have attracted a lot of attention so far. You can simply download the file to your phone or computer. So, how does such an item suddenly become so valuable that people are willing to pay millions for it?

It's human nature. The scarcity principle is a construct of social psychology where we assign a higher value to rare items, while on the other hand, we devalue those that exist in abundance, or are commonplace. To be precise, there's no guarantee that people will buy into the hype that something is scarce, unless they have tangible reasons to believe this is a fact. Therefore, proof of scarcity is necessary for this principle to hold.

This explains why some NFTs are more valuable than others. Take the example of NBA TopShot Moments. Common Moments make up the bulk of their NFTs. Tens of thousands are created and released into the market. These are some of the cheapest in that ecosystem. On the other hand, the developers only release a few hundred Legendary Moments NFTs, which are pricier. Finally, you have the Ultimate Rarity Moments, which are only released for auctions, and are the most expensive so far.

Another good example of this would be the art industry. Art valuation is generally based on the assessment of a select few individuals in the industry, who decide whether some work is valuable or not. This is what has held the art industry together for centuries. Other than that assessment, all you have in the case of a painting, for example, is canvas, paint, and a creative mind.

Demand in the NFT market has mostly been driven by hype, especially around the rarity of some NFTs. While their rarity is not in doubt, the hype pushes their prices to astronomical levels. Note that the blockchain market is mostly a speculative market, so it's also difficult to tie tangible economic factors to the valuation or devaluation of NFTs and other digital assets on the blockchain.

Given this context, value in NFTs is derived from the fact that you can verify the authenticity of the NFT. Despite the hype and speculative nature of the market, validation as supported by the native structure of the blockchain makes it reasonable for people to buy NFTs.

CHAPTER 4
THE DIFFERENT TYPES OF NFTS

t was only a couple of years ago that NFTs were unheard of. Today, they exist in so many variants, you can find NFTs representing pretty much anything in modern society. As we continue to push the limits of our creativity through NFTs, they will redefine the boundaries of modern investments and the nature of assets. This is a booming, growing market, and we can anticipate massive growth in the next few years in each of the categories we'll discuss in this chapter.

The NFT ecosystem has grown so much that we now have dedicated marketplaces springing up all over the place. Many developers have experimented with NFTs, and thanks to the vast variety in their use cases, the market for NFTs is becoming more defined over time. To help you figure out the different marketplaces and how to profit from them, let's first understand the different types of NFTs:

ARTWORK

If there's a category of NFTs that has really popularized this market, it's visual art. There are artists who are making a good living from their art NFTs. We might think of art NFTs as an emerging industry because of the boundless nature of art. Even in the traditional art world, contemporary art has evolved over the years, such that there's a very thin line between what we can consider art and that which we cannot.

NFTs allow artists to take this dimension a notch higher, especially since now artists on the blockchain have access to a global audience. The art industry tends to be a close-knit society where you have to know people in strategic places to get that major breakthrough and make good money.

NFTs have made it easier for artists at all levels of the trade to access audiences and sell their work. From exclusive auctions by big names like Sotheby's to Dartroom, there's a market and an audience for everyone. It can be argued that NFTs have helped to make the world of art more inclusive.

Coming off the economic turmoil of the pandemic, showcasing art NFTs online has helped many artists expand their reach, which has been great for small, local artists who were previously geographically limited. Even better, artists can now earn an income *immediately* after a sale is made, which is quite encouraging for the creative industry in general.

COLLECTIBLES AND TRADING CARDS

Collectors have always held trading cards in high regard. The community around trading cards is so elaborate that you'll find different kinds of annual trading card shows all over the world. NFTs has made it easier to digitize novelty projects like trading cards and collectibles, which opened up a new market for collectors.

In-game assets, for example, are no longer restricted to their inherent gaming environments only. Some games that have embraced blockchain technology allow gamers to use their digital assets throughout the gaming ecosystem. For example, if you buy in-game assets in *Axie Infinity*, you should be able to use them in *Decentraland* too. This is just an example of how NFTs are changing the dynamics of the wider collectibles market.

Ultimately, NFTs are reshaping our perception of owner- ship. In the world of collectibles, there's value in the aspect of rarity, which happens to be a key feature in assigning value to NFTs.

Together with trading cards, collectibles and NFTs are a perfect match. These items have always been valued for their exclusivity, and in some cases, the hype around owning them. This is why collectors are often willing to pay premium rates to own some items, or add them to their collection. Apart from rarity, NFT trading cards and collectibles are also held in high regard because of certifi- cation and utility value, similar to the physical components.

DOMAIN NAMES

In a market where so much money is moving online, it was only a matter of time before domain names joined the growing list of items that can be tokenized as digital assets. The most notable NFT domain sale at the time of this publication was the $100,000 sale of win.crypto, which also happened to be the most expensive domain NFT sale in history.

A domain is home to a website. A domain NFT is no different, just that it is home to a decentralized website. Another notable difference is that instead of having the traditionally complex wallet address, we have a name that's readable and easy to understand, like win.crypto or gambling.crypto.

Gaming and sports domains are among the domain NFTs that are getting as much attention as their native gaming and sports NFTs. The value in selling domain NFTs isn't just integrating domain names into the blockchain to get cryptographic addresses that are simpler, and easy to comprehend. It's also about making the purchase process for domains much clearer than they usually are, and speeding up payment and confirmation processes for customers.

The list of things making record-breaking sales as NFTs keeps growing, from digital art to domains, virtual real estate, and even love. The NFT market is a revolutionary force that's not just affecting music and the arts. It's drawing attention from everywhere.

THE DIFFERENT TYPES OF NFTS   53

EVENT TICKETS

NFTs and event tickets were made for each other. The peculiar features of NFTs easily fit into solving some of the challenges that the ticketing industry has had for years. Their non-fungibility can help address the problem of issuing multiple tickets for a seat, which usually results in chaotic scenes as each holder claims validity.

Ticket ownership and validity can also help to tackle ticketing fraud, such that the authentic ticket can only be owned by one person. If they sell the ticket, they immediately lose rights to the seat and the ticket to the new owner. This, and the integrated payment solution, makes it impossible for ticket cheats to penetrate the ticketing industry as they've done for many years.

Creating tickets on the blockchain allows event owners to build some rules into the ticket smart contracts. For example, event owners could make it impossible to transfer tickets to new owners, or claim a percentage fee each time a ticket is resold in a secondary market.

By taking control of secondary actions around tickets, it might also be easier to tackle the problem of ticket usury. Because of the authentication processes, this might also help overcome the problem of hooliganism that can invade sporting arenas.

NFT tickets could also be useful in providing more transparency to event headliners like musicians, so they know the exact number of tickets sold, instead of waiting on a third party to confirm sales. Artists and fans can look forward to a direct relationship without the need for tick-eting agents and other middlemen.

GAMING NFTS

Gaming NFTs are a blockchain concept that allows gamers to use in-game assets as real world assets like bonds and stocks, through the NFT marketplaces. Gamers don't just own their in-game assets, they also have the right to transfer or sell them to other interested gamers. Commonly traded in-game assets include gaming skins, avatars, and weapons.

Through NFTs, the gaming world has created vast oppor-tunities for all gamers to not just enjoy interacting with their favorite franchise, but also to take advantage of the entrepreneurial perks while at it. Metaverses like Sandbox are leading the charge by allowing gamers to actually buy and develop virtual real estate. It is such a big business that the Sandbox in-game token, SAND, boasts a market capi-talization of more than $4 billion as at May, 2022.

There are more than 2 million active users who interact on Sandbox every month, with some of them using the gaming metaverse as a reliable source of income. This concept has also revolutionized the commercialization of modern gaming such that it's not just the franchise owners and their partners who get to profit from the game, indi-vidual users can do so too.

Interest in NFTs doesn't seem to be slowing down, so we can look forward to more gaming franchises flocking to this scene in the future, which also translates to more income-earning opportunities for gamers. With this trend, gaming won't be the proverbial waste of time after all.

MAJOR SPORTING MOMENTS

There are sporting moments that every sports fan remembers vividly for years, the kind of moments that you grow old and fondly tell your children and grandchildren about. This is why sporting moments NFTs exist.

Major sport brands are using NFTs to enable tokenization of special game moments. It could be anything from a LeBron James dunk, to a game-winning Cristiano Ronaldo free kick, or Sergio Aguero's spectacular last-minute, season-ending goal against Queens Park Rangers in 2012 that took the title off Manchester United's grasp and handed Manchester City a win. These moments, to sports fans, are the stuff of folklore.

Other sporting activities that have since been tokenized include memorabilia and player cards. This is the next level of fan engagement and interaction, which could also help teams and the sporting industry in general tap into new revenue streams through promotions, advertisements, and supporting income-generation approaches that deliver multi-faceted value points for teams and fans alike.

While NBA TopShot has received most of the plaudits for using NFTs to make special game moments come alive forever, other brands have since come into the fold. Soccer fans, for example, can swoop to Sorare to get something quite close to the NBA TopShot experience.

MEME NFTS

It's interesting how NFTs are earning people millions of dollars, despite the fact that it was only a few years ago that most people had no idea they existed, let alone knowing what they were all about. In fact, for many people, the entire blockchain world is a new concept, and even though they know there are lots of opportunities in it, taking a leap of faith might be asking too much.

A meme is an image, video, or text (often humorous in nature) that quickly gains popularity on the internet and is spread quickly in conversations amongst users.

Memes can be fun, sarcastic, and are very often very witty. They originated simply to entertain and ride a short zeit-geist wave, but have evolved to have some serious income-earning opportunities. There's a good chance that most of the popular memes you've seen or used in the past, are now worth a great deal of money in NFT marketplaces.

It's been quite an exciting spectacle seeing how the meme NFT market has evolved. Take Doge, one of the most iconic memes out there. The meme is essentially a photo of Kabosu -a Shiba Inu dog - that was taken in 2010. It features Kabosu with an internal monologue in deliber-ately broken English, and it is often used to express anything from sarcasm to disapproval. Even though the photo existed earlier, the meme became famous in 2013 and is widely considered one of the best memes in existence.

The actual NFT of Doge sold for around $4 million. The sale sparked so much discussion in the NFT community that it even resulted in a cryptocurrency being created from it. And to think it all started with an innocuous photo of someone's pet.

Other notable meme NFTs include Bad Luck Brian, Disaster Girl, Charlie Bit My Finger, and Nyan Cat. It's been quite a fascinating experience seeing the advancement in meme NFTs over the years, even though it's not as easy to monetize memes as many people think. The famous Charlie bit my finger video sold in 2021 for approx $750,000 , and despite the extravagant figure, it's a lot less than some experts believe it would sell for.

MUSIC AND MEDIA

With celebrities flocking the NFT scene, it was always just a matter of time until the music industry caught on. There have been lots of exciting NFT projects in the music scene since 2020. Because of the pandemic, many musicians were grounded, and could no longer rely on outlets like concerts and live shows for income. Merchandise sales also slowed down, and as time went by, things only got worse. By 2021, the music industry had learned from the rest of the world, and many musicians started working on alternative methods of earning through NFTs.

Just like the visual art industry has felt the impact of NFTs, the music scene will equally experience revolutionary changes over the years. This is likely to happen as music is an industry that thrives on communities and loyal fans. Because of this community aspect, musicians who leverage the use of NFTs properly will continue to forge stronger bonds with their fans.

This is an exciting time for both musicians and their fans as the industry explores a different dimension of entertainment. Rock band Kings of Leon, for example, made history as the first music group to release an NFT album. For their fans, who had been waiting three years for the rock band's latest album, it wasn't *just* the music that excited them. To have the band release it in this manner, and the timing of it, was in itself a work of art, and the move paid off, handsomely.

Major media outlets and publishers like CNN, The Atlantic, and Associated Press also started using NFTs to shore up their revenue streams during the pandemic. Media and music have always been early adopters of emerging technologies, mostly to keep up with the demands of their audiences. And in adopting NFTs, they kept up this trend. It was also inevitable perhaps, and can be seen as a natural step in the progression of the two industries.

REAL ASSETS

Apart from digital assets and tech products, NFTs have also been used to represent tangible assets in the physical world. The best example of this is in the real estate sector, where NFTs have helped to streamline the contractual process by eliminating unnecessary paperwork and middlemen. From the real estate industry, it's evident that the use of NFTs is likely to prevail in other sectors too, especially where proof of authenticity and ownership is necessary.

Real asset NFTs are a bit more complex to implement in practice than they are in theory. In real estate, for example, you can tokenize the entire asset or use NFTs to issue fractional ownership. Fractional ownership simply means allowing users to own bits of an asset, more like what happens when you buy shares in a company. In this case, everyone receives a number of tokens whose value represents their ownership in the house or real estate holding.

Tokenizing an entire house, on the other hand, involves creating an NFT of the title or deed to the house. Even though creative real estate agents and companies have managed to create these NFTs, the appropriate regulatory frameworks are mostly still non-existent, or unclear on the dynamics of real estate NFTs. While this might be an exciting concept, most people might feel uneasy about the absence of regulation, especially in a digital financial market where things are volatile and can change all the time.

On the flip side, digital real estate has taken off in virtual environments like metaverses. This is mostly because users have the freedom to create their virtual realities. Virtual real estate is thriving in metaverses like Roblox, Minecraft, Decentraland, Somnium Space, and Axie Infinity.

One of the most promising aspects of adopting real estate NFTs is in the purchase process. Smart contracts are replacing middlemen. This will eventually make the ownership transfer process faster, in addition to the fact that all information on ownership of the property is written immutably on the blockchain.

VIRTUAL FASHION

Like gaming and the arts, the fashion world has also picked up on NFTs. Auction houses like Christie's have hosted several successful auctions for brands like Gucci, and the overnight success has seen several fashion houses getting on the Ethereum blockchain. NFTs have helped open up the fashion scene to a new set of audiences, especially younger, tech-savvy customers.

Their emphasis is to try and capture the attention of new audiences by offering value beyond what's already available in the traditional fashion market. Through NFTs, luxury fashion brands are encouraging audiences to interact with their brands in a different environment that fosters exploration, art, and design.

Luxury fashion's entry into the NFT space was always imminent as a natural step in the progression of NFTs, since fans were already thrilled about using skins and other kinds of digital fashion. NFTs offer the one thing that fashion fans have always struggled with, authenticity and exclusivity.

From the Nike patenting CryptoKicks example we mentioned earlier, we can clearly see that luxury brands are taking this development seriously. The concept of tokenized shoes works in such a way that once a customer buys the shoe, they unlock the token using an identification code that ultimately links them to the shoe on the blockchain. The digital representation is a welcome move that might help fight the counterfeiting scourge that has plagued the industry for years.

As high fashion gets serious about NFTs, we'll see more development in legislation around this space. For example, as a result of their patent, Nike made headlines by filing a lawsuit against StockX, a sneaker and apparel exchange, for infringing on their NFTs by selling NFTs using Nike's trademarks without their approval or authorization.

Luxury fashion brand Hermès also filed a suit against a digital artist for appropriating their brand online through his NFTs. For major outlets to get this concerned about their patents is proof of their realization of value in blockchain assets like NFTs.

CHAPTER 5
HOW TO MAKE MONEY FROM NFTS

There's undoubted value in NFTs. And that should be a valid blanket explanation for why people are spending so much money on them. Breaking it down however, demand in the NFT market can be attributed to a lot of different things, including our innate desire to own the next big thing, distrust in the stability of currencies, or just to keep up with trends.

Art has for a long time been used as a store of value. It's easy to see how the blockchain furthers this concept. NFTs have opened up art spaces to more audiences from all walks of life. In effect, this created room for artists to make money beyond traditional audiences. Not very many artists have profited from the mainstream market with exclusive auctions and so on. The average artist mostly splits their time between art and something else, perhaps a full-time job, to get by.

As a store of value, NFTs have made it easier to use art in the same way we use precious commodities like Bitcoin and gold.

Buoyed by a booming tech industry and younger generations more receptive to disruptive technology, people have different motivations to buy NFTs. A prime motivator is the unique connection buyers have to the content creator, which tends not to exist with traditional art forms of the past.

Furthermore, demand for NFTs can be explained in simple economic concepts: scarcity, utility value, and the unique connection to the creator that comes from the ownership history, which we highlighted in Chapter 3.

SETTING UP YOUR ACCOUNT

Before you think of making money by selling NFTs, you must create a digital wallet. There are different kinds of wallets available, but at this point, I'd strongly recommend MetaMask. Most of the major NFT platforms you'll come across use MetaMask. Given how close-knit communities are around NFT markets, if the majority are using Meta-Mask, that's a good enough reason for you to consider it as a starting point.

Visit metamask.io to set up your account.

You'll need to download the Chrome or Firefox extension to your browser. Even though you can still install it on your phone, it's slightly easier to interact with NFTs on your computer. The user experience is currently better on desktop devices than mobile environments.

Once the extension is added onto your browser, click on Get Started and follow prompts to set up and customize your wallet.

Create your password and recovery phrase. Your recovery phrase is useful for account recovery purposes, so just like your password, you must keep it safe. If someone has access to your recovery phrase, they could easily take over your account and steal your funds, then terminate your account altogether.

Losing your recovery phrase might lock you out of your account forever. To avoid that, write it down and store it in a safe, offline place.

MetaMask is hosted on the Ethereum blockchain. Therefore, you generally won't need to switch to another blockchain to use it. You can also go to your account profile and edit the details as you see fit.

You'll find a QR code you can share with other users in case they ever need to transfer some money into your account. The QR code links them to your public address, which is how you receive money.

All transactions conducted through your MetaMask account will be displayed under the activity tab. This is also where you see the total amount available in your Ethereum account listed under Assets.

COMMUNITY VALUE IN NFTS

NFTs have advanced from hype projects to utility value projects because of the communities around them. If anything, many people buy into NFT projects, not necessarily because of the artwork or whatever asset is behind the NFT, but because of the fact that they are joining a community of like-minded persons, excited about advancing the NFT project beyond the confines of a digital asset. Some of the benefits that come from joining NFT communities include physical access to exclusive events and enlightening interactive sessions on social media platforms like Discord and Telegram.

Gary Vee's The Flyfish Club (FFC) is an example of what it means to buy into a community, and not the project. The exclusive members-only private dining club gives NFT holders access to world class social, cultural, and culinary experiences through the restaurant. NFT holders are also allowed unlimited access to a private dining experience in an iconic setting in New York, complete with an outdoor space, intimate omakase room, upscale restaurant, and an amazing cocktail lounge.

We also have Loot, another exciting NFT project built around its community of users. What makes Loot and many other NFT projects that have established communities unique is that while the idea could be replicated and used in other projects, the communities around them are not.

The Loot concept is all about simplicity, making it open to value in both the primary markets and derivative secondary markets. The project appeals to your creativity and imagination, and arguably one of the projects whose dimension is aligned with the general concept of a metaverse.

A community represents the individuals who hold tokens in an NFT project. We can further expand the definition to include those engaging in the project before its official release, especially on social media, or anyone interested in minting the NFTs.

Communities are integral to the success of NFT projects, hence the need for you to research them well before you buy into any NFT. Even as you research the nature of NFT communities, you must be careful, since some projects hire influencers to push their agenda. This was especially evident in the summer of 2022 when renowned English footballer Michael Owen tweeted "My NFTs will be the first ever that can't lose their initial value." He was forced to delete his tweet by the Advertising Standards Authority as they ruled that tweet misleading to consumers who were likely to trust the English celebrity.

While there's nothing wrong with creating hype around a project, some influencers barely understand the technicalities of the projects they're promoting. The problem here is that you could end up buying into a project with more casual users than development-minded users. Therefore, just because a project has a lot of users on social media doesn't necessarily mean that the community around it is strong enough to support and grow it.

A strong and resilient community around an NFT project is generally the result of active participation, frequent interactions, and more importantly, users who hold a long-term view of the project at hand. If possible, try to follow the conversations and discussions in a project community to gauge the direction of their commitment to the project.

As NFT projects go through different growth stages, their inherent communities could also evolve and become more exclusive. This is common as soon as minting is done, which locks out the members who were only active on social media, but never purchased any NFTs. The reasoning behind this is that since they don't hold the tokens, there's a high chance that their contribution to the project might not be relevant over the years. Without the tokens, such users could possibly be inactive on the project, or move on to other projects in the future. Therefore, token ownership is considered a statement of intent, hence the move to exclusivity.

Beyond the minting phase, community members usually own the project. They believe in the project, and for that reason, engage in constructive discussions about how to improve the project.

Before you buy into any NFT project, you should always try to learn as much as you can about the community around it. The general consensus is to avoid projects that are shrouded in anonymity, or those where the mood around the community is negative. If there have been lots of complaints about the project within its community for a while, there's a good chance that position might not change. Such wouldn't be a good investment opportunity.

When assessing the nature of the community around an NFT project, you should also consider whether there's reason to believe the project will be sustainable over the long term. Longevity and the general emotion in the community should give you a good feeling of whether it would be a good investment or not.

Try to get into their Discord and Telegram channels to get a feel for the conversations. Since community members are mostly people who own the projects, consider how they treat beginners, or anyone who asks beginner-level questions. If they are welcoming and willing to help beginners learn at their own pace, that's a sign of an awesome community. On the other hand, if they ignore or brush you off, you really don't need that much negativity in your life, especially for a project you are about to pump your hard-earned money into.

Pay attention to mentions or concerns of a quick flip in the conversations. If the community members are more concerned about making a quick profit, it's highly likely that they believe the project will only be valuable over the short term.

One of the essential roles of a community project is to create and sustain value in the project. Note, however, that membership in an exclusive communication platform like Telegram or Discord doesn't necessarily guarantee value or tangible utility in the NFT project. As much as it might be fun to interact with other users, you should never mistake the engagement for utility. Besides, once you've been in the game for a while, you'll realize that after purchasing or selling a few NFTs, you'll barely have the time to fully participate in the community engagements. This is a normal experience many people go through. So, even as you research the communities around potential NFT projects you'd love to invest in, keep an open mind.

Finally, another telling feature of a good NFT project is the connection between the founders and the rest of the community. How often do the founders pop in to engage in discussions, provide clarity, or any other issue that arises? More importantly, do they discuss the roadmap, their future plans, and goals for the project? This is important because people buy into the project as a result of their plans for the future. This is actually one of the crucial features of the project that might assure stakeholders of sustainability and value over the long-term. If that is assured, the community will actively buy and hold onto the project, as opposed to those who buy to flip for a quick profit.

WATCH THIS SPACE

With millions of NFTs for sale in a global marketplace worth trillions of dollars, a breakthrough is always in the offing for someone. Such is the beauty of the internet. You never know when something mundane might become the next big thing. Here are some exciting NFT projects that might possibly have an amazing future in 2023 and beyond:

- **Dribblie**

This is an exciting football manager game that involves an inter-galactic battle for glory. NFTs are used in-game as stakes, leased or owned assets, allowing users to earn by playing football. There are several strategies built into the platform, which essentially allow you to earn up to $100 a day from gaming. Essentially, you earn money by winning games, trophies or quests.

- **Azuki**

This is a skater project that integrates the best of the digital and physical worlds. The 10,000-avatar project allows you access to different dimensions, such as the garden, which is the point of confluence between the digital and physical world. There are plans to introduce live events through Azuki in the future, which will make the project even more surreal for users.

- **Doodles**

This project is built off the work of Burnt Toast, with 10,000 NFTs each with unique visual features. The doodles include mascots, apes, aliens, cats, and skellys. The entire collection will also include costumes, rare heads, and other creative works depicting the author's artistic palette.

Doodle ownership gives users the benefit of voting on future events and product releases, and any other initiatives supported by the community. The Doodle project brings together designers, project architects, and collectors, in what should grow into an incredible collectors' haven.

- **Women Rise**

This is an exciting NFT project whose timing couldn't have been any better. The struggle for inclusivity and diversity is incredibly important in all areas of our society, so a project bringing this cause to the blockchain will be one worth looking out for.

Women Rise builds on the work of Maliha Abidi, an internationally recognized visual artist, with 10,000 random collections aimed at introducing more female characters into the NFT marketplace. Other than that, some of the sales proceeds will go towards supporting different gender equality initiatives like advocating for mental health awareness, and supporting girls' education in underprivileged communities.

- **Autograph.io**

If you love sports, this project might interest you. While it's presented as an exciting NFT project, its definitive features make it more of a marketplace with a community built around special or unique digital moments, cultural experiences, entertainment, and sports. Co-founded by Tom Brady, Autograph.io boasts a board of advisors including the likes of Tiger Woods, Wayne Gretzky, Usain Bolt, Eva Longoria, Naomi Osaka, among other famous personalities in the entertainment and business world.

From the onset, the project maintained an athlete-oriented approach, promising private Discord channel entries, access to future drops, and authentic signatures from featured athletes. However, given the personalities involved, the project has the potential to diversify into the wider sport scene, and probably entertainment and celebrity segments.

What we know for a fact is that it's not easy to figure out whether an NFT project will be a worthwhile investment or not. When choosing a project to invest in, you should strongly consider the utility value of the project. Utility is one of the primary reasons why people consider investing in a project. What value does the project offer the community around it? The pursuit of utility is also the reason why many marketplaces and projects are being created: to offer something more than just the artwork, like many of the NFTs that have been in the headlines for so long.

To be precise, some artwork NFTs do have utility value, especially those that offer benefits like exclusive memberships, access to events, and so on. As you research the market, it would be wise to choose a project that can benefit you in some way, and not just a hype project.

Please note that the content in this section is for educational purposes only, and not financial advice. Conduct your own research and with due diligence, use the information in this book at your own risk.

CHAPTER 6
NFT MARKETPLACES

NFT marketplaces are platforms where users trade in NFTs, similar to stock exchanges. You can also think of NFT marketplaces as the eBay or Amazon of NFTs. Apart from buying and selling NFTs, marketplaces also provide an opportunity for users to store, display and in some instances, even create NFTs. Below are three important things you must have before you can access an NFT marketplace:

- **A digital wallet**

There are different kinds of wallets, so you should choose one that can be used on the blockchain on which the NFTs you want are hosted. Naturally, since most NFTs are hosted on the Ethereum blockchain, it would be wise to start with a wallet built on the Ethereum platform, like MetaMask. On the other hand, if you want to buy NFTs built on the Solana platform, you'd need a wallet compatible with Solana, like Sollet.

- **Some crypto**

You must have some cryptocurrency (usually Ether) in your wallet before you list, buy, or mint an NFT. Similar to the example on digital wallets, find out the currencies that are accepted in a marketplace before you sign up. Luckily, most marketplaces support the major cryptocurrencies.

- **Trading account**

Finally, you must set up a trading account on the marketplace. This is where all your market transactions will take place.

We should also point out that most marketplace transactions attract a fee (gas fee). The cost varies from one blockchain to the next, and other inherent factors, for example, the level of activity on the blockchain at that particular moment. Gas fees are generally higher when the market is overcrowded, so it's wise to study the level of activity on a marketplace for a while, to figure out the most appropriate times during which you can spend less on transaction costs.

Popularity is also another factor that determines the cost of gas. Being the most popular and hosting the largest number of decentralized apps, Ethereum is often the most expensive blockchain in terms of gas fees.

There are several marketplaces available today, catering to diverse user needs. This is a good thing, given that the wider NFT ecosystem is still developing, and will probably evolve in different dimensions over the years.

CHOOSING THE RIGHT MARKETPLACE

As there are different kinds of marketplaces in the blockchain ecosystem, it's important to choose one that aligns with your NFT needs, and any other requirement you might have. The digital asset environment is still evolving, and as more projects detach from the dominant Bitcoin and Ethereum, marketplaces will similarly evolve and adjust to the unique project needs. Below is a simple guide to help you select a marketplace that is appropriate for your NFT needs:

- **Type of Projects**

The consideration at this point is whether you are creating multiple NFT projects or just a single project. Some marketplaces only support single projects while others allow you to create multiple. Some of the big marketplaces that restrict you to a single project include CryptoPunks, Axie Infinity, and Mobox.

Your project type will also weigh in on the cost of using the marketplace. As we mentioned earlier, this is an important factor that will affect your profitability online. For a beginner, or if you have been around for a while, but your budget is limited, you'd be better off avoiding marketplaces hosted on the Ethereum blockchain, which generally charge high gas fees.

- **Market Information**

Everything you'll do in an NFT marketplace is an investment. Investments and information always go hand in hand. Choose a marketplace that provides as much information as possible about the projects listed on it. For example, you should be able to see, apart from the price of the NFT, the number of listings available, number of holders, the buying and selling activity on that NFT, and all relevant historical trading data to help you make a decision.

- **Network Security**

When it comes to your investments, you cannot compromise on security. A good marketplace should have appropriate measures to protect buyers as well as the creators on their platform. You might not know much about the technical aspects of their security protocols, so the best thing to do is research.

Find out whether the marketplace has ever been hacked before, the magnitude of loss, and how they handled the situation. You can also follow conversations in different forums about the platform's security protocols. Some basic expectations for secure marketplaces include insuring transactions to avert fraud, insistence on identity verification, and encrypting private user data.

- **Informative Search Capability**

Whether you are buying NFTs, selling, or just researching the market, it's important to have access to reliable data. An ideal marketplace should support in-depth search functions, providing clear visual data so that it's easier for you to navigate the market.

As there are many NFTs being added to the marketplace regularly, the platform's search functions should make it easier to assess listings at a glance, helping you figure out whether a project is worth investing in or not. For example, you should be able to search using the creators' names, types of NFTs, or even a specific named NFT.

- **Simplicity and Convenience**

Every decent marketplace should be secure yet simple enough for all users regardless of the device or location. If anything, you should have convenient access to the marketplace on demand. This also means choosing a marketplace that can support multiple payment alternatives and NFT wallets. This makes your work easier such that your payment options are not limited.

- **User Experience**

When you're using a marketplace for the first time, it's smart to know what other people think about their services. This is where reviews and ratings come in handy. The last thing you want is to be the guinea pig for a marketplace, while you are spending so much of your money. A simple internet search should provide most of the information you need. However, if you want the best, relatable reviews, check the community forums built around the marketplace.

With that information, you should be able to choose the most relevant marketplace for your NFT project needs.

Remember that if you are creating NFTs to sell, this is the experience that your audience will have. Therefore, if you face challenges with a marketplace, there's a good chance your potential customers will equally struggle, losing you potential income in the process.

The popularity of NFTs will continue to rise in the future as we come across more mainstream uses for the digital assets. With that in mind, it's ever so important that you learn how to choose the right marketplace so that you can maximize your chances of getting high returns on your investment.

COMPARING YOUR OPTIONS

There are quite a number of marketplaces for NFTs available, and many others will probably come up in the future, which begs the question: which is the most ideal marketplace for your NFT needs? We already discussed some factors you should consider when choosing marketplaces, so let's take things a notch higher, and review some of the top marketplaces in the industry in terms of what they have to offer.

Remember that as much as you are looking for the perfect marketplace for your NFT needs, you are also looking for a cost-effective solution that meets both your needs and those of your customers. Cost, in terms of gas fees, is of particular emphasis here. To make your work much easier, we will also classify the marketplaces into two categories, open and closed markets, and discuss ease of access.

OPEN MARKETPLACES

An open NFT marketplace is one where entry is free for all NFT creators. This ease of entry makes open marketplaces an ideal place for most creators to start. You'll also interact with more content here, so it's also a good place to start and establish a name for yourself before you move on to exclusive marketplaces.

OPENSEA

OpenSea has been around since 2017, and is widely considered the Google of NFT marketplaces. Almost every NFT investor goes to OpenSea for their NFT projects. Given its large population of registered creators, many beginners also consider OpenSea to get first-hand experience of the ins and outs of the NFT business.

One of the benefits of using OpenSea is that it supports a variety of file formats, which means most of the media you can think will be supported. Other than the usual suspects like jpgs, video, and GIF files, OpenSea also supports other types of physical and non-physical assets.

On your first minting contract on OpenSea, you will have to pay two fees before your first NFT sale. The first one sets up your account, and costs anywhere between $70 and $300. The second fee allows OpenSea to access your NFT, and costs between $10 and $30.

Once you make these payments, you won't be asked for gas fees from your second minting contract onwards. Therefore, as much as the initial fees might be expensive, the structure for OpenSea makes it a relatively better place to be for someone who wants to invest in NFTs for the long-term. You'll only be charged a 2.5% commission on sales.

The flip side of using OpenSea is that the market is already flooded with creators, so for a beginner, your listing will be competing for audience attention against the established creators, and everyone else—literally millions of other listings. Therefore, you won't get as much exposure as you'd get in a closed marketplace.

Considering the challenges in getting exposure, the best way to work with OpenSea is to promote your work on social media, and only use the platform for sales.

RARIBLE

Rarible goes toe to toe with OpenSea in terms of popularity, even though it's only been around since 2020. It is a reliable marketplace and boasts a large user database like OpenSea. It supports minting NFTs on the ERC721 standard as well as its native token, RARI. It features multi-format support, so you can list your NFTs as audio, video, or image files.

One feature that sets Rarible apart from other marketplaces is that its native token, RARI can be used for other purposes apart from trading NFTs, mostly to vote on proposals, or to make proposals for changes in the marketplace. Common reasons for voting include supporting its community members through grants or moderating members. The token allows users participatory rights similar to what you'd expect when you buy stock in a company.

Similar to OpenSea, you'll first pay the initial gas fee when mining your first contract, but you will incur listing fees each time you sell an NFT. Towards the end of 2021, Rarible introduced gasless minting, which is ideal for creators who are wary of gas fees, or if you are just getting started, and can't afford to pay upfront gas fees. Note, however, that like other marketplaces, Rarible activities will still attract a service fee of 2.5%

One of the reasons why users flock to Rarible is because the marketplace was built upon the success of OpenSea. Therefore, there are significant improvements on its core functionalities, particularly the user interface that is friendly to both creators and collectors. Other than that, creators also enjoy the benefit of being discoverable, which makes it a good platform even for beginners.

HIC ET NUNC

This is an interesting case of a marketplace whose fate hangs in the balance. As far as the internet is concerned, Hic Et Nunc was discontinued after the founder decided to pull the plug on the project. However, since all the NFTs were held on the blockchain, the owner's decision to kill the project didn't result in creators losing their work. The community around the Hic Et Nunc project created a mirror site to sort of continue the project and allow creators a chance to keep earning money from their work.

It's still unclear whether the owner might decide to revive the project or not, but there have been discussions on transitioning the marketplace into a DAO, giving creators control over it.

Challenges aside, this was a popular marketplace for creators seeking affordable alternatives to the high gas fees on the Ethereum blockchain. NFTs on Hic Et Nunc were traded in Tezos (XTZ). The minting commission paid to Hic Et Nunc was 2.5% before the marketplace was discontinued, but has since dropped to 1% while the marketplace operates on a mirror website, making it one of the most affordable NFT marketplaces. Hic Et Nunc's ability to offer low gas fees comes down to Tezos' low energy consumption compared to Ether, for handling NFT operations.

While this marketplace supports different NFT formats, its downside is that Tezos isn't really a popular blockchain yet.

MINTABLE

Aside from the two major marketplaces—Rarible and OpenSea—Mintable comes in as a great alternative if you are concerned about gas fees. On Mintable, you can mint and list your NFTs for sale without paying gas fees. This is the gasless method, one of two options you get when listing a project on Mintable.

The other is the traditional approach, which means you will mint the NFT, but have to pay some gas fees. If you choose the gasless method, you won't incur any network fee when minting your NFT. However, your NFT will only show up in your wallet once someone buys it.

Note that in the gasless method, Mintable will still charge you a 5% service fee, though you will only pay for it once your NFT has been sold.

While Mintable sounds like a cost-effective marketplace for beginner creators, they only support 3D files, video, and image NFTs. If you intend to create audio NFTs, this might not be the most ideal marketplace for you. However, the general workaround for this is to format your audio track by attaching a video or image file to it, and uploading in MP4 format.

ZORA

Zora is another exciting platform that is poised for future growth, with interesting features for both collectors and creators. You can mint contracts using both the ERC20 or ERC721 network. This flexibility is also the reason why it supports NFT transactions in different currencies.

Zora might be a good place for you to get started, especially since there are no service fees involved at the moment. However, you'll still have to pay gas fees. If you market your work properly, Zora can be a great marketplace especially since they won't charge you any commissions.

Another good thing about using Zora is that there's in-built support for different file formats, including text, illustrator and Adobe PhotoShop files. Unfortunately, it is still an up-

and-coming marketplace, so it doesn't have as many collectors as the established competitors.

CLOSED MARKETPLACES

Open marketplaces are a good place for beginners, mostly because of the large number of collectors and audiences, and the fact that they can allow you some incentives like free listing. However, once you've been in the industry for a while, it might be great for your brand if you could diversify and explore other options in closed markets.

You can think of closed marketplaces as exclusive or premium networks that were created to serve specific needs, as compared to the open marketplaces that are free for all. Most closed marketplaces have stringent entry requirements that you must meet for your membership to be considered. Furthermore, you can only get in through an invitation by one of the existing users.

DARTROOM

Dartroom is an NFT marketplace that allows creators to list NFTs either as Algorand standard assets (ASA) or as smart NFTs. Since Algorand is considerably cheaper to mint compared to other digital currencies, Dartroom is positioned to attract many beginner artists and creators who might have been priced out of other marketplaces.

Dartroom also banks on the promise of a forkless Algorand project, thus creators don't have to worry about duplication in the market. While the NFT market thrives off the benefits of blockchain technology, it is also susceptible to any risks inherent to the blockchain. One of these is a hard fork.

A hard fork basically results in two versions of the blockchain, one using the old rules, while the second one uses the new rules. The challenge here is that NFTs that exist on the old blockchain will be duplicated on the new blockchain, defeating the essence of originality and authenticity of NFTs.

With the backing of the Algorand project, NFT creators on Dartroom have peace of mind knowing that their tokens will never suffer the fate of forking, so they'll never be duplicated. Because of the forkless blockchain ecosystem, creators on Dartroom can also look forward to using a truly decentralized, secure, and reliable marketplace.

On top of that, Dartroom built royalties into the smart NFTs on the marketplace, with the promise that the setup will continually get improvements and updates over time. Ultimately, Dartroom's approach is to become the go-to marketplace for all creators, including those who are not so tech-savvy, or those who don't understand the workings of blockchain technology.

FOUNDATION

Of all the closed marketplaces, Foundation stands tall as one whose entry process is relatively the easiest to accomplish for new members. It also boasts thousands of creators. From time to time invitation codes are shared on social media with artists whose work is unique, or could also be exchanged for NFTs or money, which might explain why it's quite decently populated.

The community initiative behind this marketplace is also another reason for its stellar performance in terms of making it an awesome place for artists to list their work. Through their community upvote initiative, creators generally vote for one another, with the top 50 artists winning the right to join the Foundation ecosystem as creators. Being a creator on Foundation carries some clout in the world of art, because everyone knows the process of getting in is about merit.

Creators registered on Foundation generally command some respect in marketplaces because they are perceived to have earned that right by virtue of peer-reviewed or recommended high value work.

Compared to open markets, Foundation charges 15% commission, not mentioning the high listing, minting and auction settlement fees. That notwithstanding, the Foundation creator tag gives you the kind of exposure no money can buy, which is why many creators who trade on Foundation don't mind it.

The structure of the Foundation marketplace is rather basic, and is relatively difficult for new creators to be discovered. On the other hand, established Foundation creators could offer you a helping hand by featuring your work. This makes you more discoverable, which is a good thing, considering that the marketplace doesn't have a dedicated search engine for keywords, so it's impossible for other users to discover your work through organic means.

Like all the other marketplaces, your best solution is to promote your work through social media and any other form of marketing available to you. Also note that Foundation only supports MP4, PNG, and JPG files so far, though there are plans to introduce other file formats like 3D in the future.

MAKERSPLACE

Compared to Foundation above, Makersplace is a closed marketplace that's considerably more difficult to access. Makersplace is not a community-driven marketplace, so you can only get invited by approval once you submit your work portfolio and profile as an audition. You can only get an invite once you pass the vetting by the team running the marketplace, and immediately start minting your NFTs.

Given the number of applications to be vetted, you might have to wait a few months to get your feedback on the application process. In the meantime, it would be wise to try your luck with other closed marketplaces like Foundation, so that you don't miss out on the opportunities available.

In terms of fees, Makersplace is considerably more affordable than Foundation. The marketplace charges 15% commission on transactions, but minting fees are only charged once for every minting activity. At the moment, Makersplace only supports MP4, TIFF, PNG, JPEG, and GIF.

NIFTY GATEWAY

Nifty Gateway is more than your average closed marketplace, it's been in the headlines for some of the record-breaking NFT news. This is where Beeple made headlines. Other notable personalities who've made newsworthy sales on Nifty Gateway include Calvin Harris and Michael Kagan. Entry is by filling in the request form, then you have to wait for the audition process. Note that Nifty Gateway is considered an ultra premium marketplace, so entry can be quite difficult.

At the time of this publication, Nifty Gateway had increased their service fee to 20%. This might be steep, but if you consider the fact that your creative work will be rubbing shoulders with some legendary creators in the NFT ecosystem, it might be worth a try. Besides, given their celebrity appeal and their audience, your big break might not be too far away. You can mint video files in MP4 format, audio files in MP3, or in PNG, GIF, and JPEG.

SUPERRARE

This PIXURA Inc project started off as an open marketplace in 2018. However, as the NFT industry evolved, SuperRare gradually became popular, and transitioned into a closed marketplace. It is currently a closed, exclusive community of NFT creators.

Like other closed marketplaces, you must submit an application for consideration, including a portfolio, and your profile information. On top of that, you must also send in a one-minute clip introducing yourself as a creator, and the kind of projects you make.

The application review process takes time, in some cases more than two months after you sent in your application, which is another reason why SuperRare is an ultra-premium marketplace. The stringent application process also means that getting accepted into SuperRare earns you some clout, so be sure to milk it for all it's worth.

You will be charged 15% commission for each sale, which is similar to what other closed marketplaces charge. Apart from that, you will also pay a 3% transaction fee. The file formats supported on SuperRare include MP4, GIF, PNG, and JPEG.

CHAPTER 7

HOW TO CREATE SUCCESSFUL NFT ART WITH A THRIVING COMMUNITY

The NFT market is buzzing with income-earning opportunities. It's not just the likes of Beeple's playground anymore. There's something in it for you too. There are lots of ways to profit from this market. As a creative with lots of digital works on your computer, we'll discuss the simplest approach to get you started right away.

To get started, there's no limit to what you can present as NFT art. It's all about creating something that people will find valuable. You can even create an NFT game and sell it online. Once your artwork is ready, create an external link to it. This could be a website, a blog, a public page on any of your social media platforms. The link acts as a landing page, providing more details about your art.

Provide a comprehensive description such that no one can confuse your work with anything else. Use the description to show your audience what's so unique about your work, and show them why it is valuable to them. Once that is done, all you have to do is set up your NFT. You could upload a single NFT or a collection, whichever works for you. Note that most people find more value in buying collections.

Once your setup is ready, you can tweak it further and add more properties, like unlockable content, levels, or anything else that you feel might be valuable. Properties tell your audience more about the NFT, so add as many as you feel is necessary.

Next, indicate how many of the NFTs are available. This takes us back to the conversation we had in Chapter 3 about the psychological approach of determining the value of an item due to its scarcity. A unique work is probably worth more than several of the same kind. However, that's only a guideline, as many artists have sold their works of art even if more of them were in circulation. Note that even if you make a lot of pieces, they all have unique identities on the blockchain.

Finally, decide the blockchain you want to use for your art. The Ethereum blockchain is the go-to platform for most NFTs, so it should be your best option. Consider the cost of gas fees, and the speed of the blockchain. If you opt for a gas-free blockchain, make sure it is fast enough. Just like that, your NFT is now ready for audiences. Click on Create and it will be published.

BUILDING A PROFITABLE NFT BRAND

You can create digital art and list it online, or you could also use this opportunity to build a profitable brand that will make you lots of money. The sound of more money in your account makes more sense, so let's discuss how to make your way into the world of profitable crypto art.

- **Setting up your wallet**

You'll need a crypto wallet to handle your digital assets. This is also where you exchange one cryptocurrency for another. NFTs are synonymous with the Ethereum blockchain, so you'll need some Ether to get started.

- **Choosing an NFT store**

For this example, we'll use OpenSea.io as our NFT store. There are, however, two different kinds of NFT stores you can consider. You can either choose general stores like OpenSea and Rarible, which are the most popular, or if you are looking for exclusivity, you can go for curated stores.

The entry process in curated stores is a bit restricted in that you can only list your work once your request for membership is accepted. The need for restricted membership comes from the fact that there have been many fraudulent NFT projects in the past that have exploited many unsuspecting users. To avoid this, curated stores like Nifty Gateway and SuperRare will only approve a formal application upon in-depth screening.

For a beginner, curated stores might be difficult to get into, especially since you barely have any recognizable presence on social networks. Therefore, it's safer to start with the general stores, then work your way up into exclusivity.

A worthy mention at this juncture is that from the onset, you can already see an element of a social community in the offing. The vetting process helps to weed out unsavory characters from the marketplace, leaving behind legitimate artists. This is also good for the evolution and development of the art scene, seeing that you end up with a marketplace full of artists, discussing challenges unique to their profession, and suggesting viable solutions. We'll dive deeper into the community concept later in this chapter and the next, to understand how it's reshaping the future of NFTs.

Create an account at OpenSea, sign in and click Create. You will be prompted to create a digital signature. This is what you'll use each time you want to create a new listing on the store.

- **Niche positioning**

Art is a wide dimension, so it's important that you figure out what your art is about before you proceed. The idea here is to figure out your style. You can't be a jack of all trades, particularly not in a digital art environment where people share so much information.

Figuring out what you are good at will also help you align yourself with communities that can help you advance your craft online. As there's so much learning taking place in NFT communities, the best thing to do is try to fit into the right spaces, and with the right crowds.

Some artists are so good at their craft that art enthusiasts and fanatics can identify their work at a glance. That's the kind of position you should aspire to attain in this industry.

This also makes it difficult for other artists to pass off your work as theirs.

If you haven't quite settled on the kind of art you wish to sell online, don't worry. There's a learning curve, and your best approach is to join an online community of artists. Connecting with other artists can help you learn more about your skills, talents, interests, and so on. Sometimes the nudge you need is to simply interact with other people in their element, eventually bringing out the best in you.

ClubHouse, Discord, and Twitter are some of the best places where you can interact with artists and art enthusiasts. You can also check popular listings on NFT marketplaces, follow the artists and use their work for your benchmarking research.

- **Brand visibility**

Presumably you really believe in your work, so let people know about it too. Art is a borderless, boundless craft where anything goes. It's all about using your creativity to appeal to someone's interests. Since the NFT explosion of 2021, many people are flooding marketplaces to try and earn an income from art. Unfortunately, you face stiff competition for audiences' attention from artists and non-artists alike.

Think big. You are creating a brand, so act like it. Market yourself, shamelessly! Don't take anything for granted. Besides, marketplaces like OpenSea are for general audiences, so you will be competing for audience attention with non-art NFTs too.

Find out the most appropriate social media platforms for your kind of work and promote it there. Join relevant groups and promote your work. More importantly, try to be active in the communities you join, but be careful not to overplay your hand.

Rise above the amateur level and think of yourself as a corporate entity. Create a professional portfolio. Set up a website for your work. These extra steps go a long way in adding a touch of credibility to your work. Even if someone realizes you are new to the world of art, they might still be interested in working with you because of the professional appeal.

Don't forget to brush up on your art personality too. Your work might get the attention of renowned collectors. Most collectors don't just buy your work, they also want to know your story. They want the full experience, and this is where your personality comes in.

- **Consistency**

The attention notwithstanding, NFT art is an emerging industry. There are many unknowns, especially for beginners. While some artists have made fortunes from NFT sales, others have struggled, and that's normal. Keep your head down and do the hard work. You might not strike gold at your first attempt, but that shouldn't make you give up. There's so much room for growth.

Almost every other artist who sets up their NFT store online hopes to become the next Beeple, but that's just wishful thinking - be realistic. Push your limits, deliver your best work, promote it, and network with other artists. As a first-time artist, there's no shame in admitting that you don't know something, and asking for help.

Luckily for you, most NFT communities online are quite welcoming, and you'll come across many people willing to hold your hand as you learn the ropes. At this juncture, the most important thing is to maintain your resolve and never give up. Keep working on improving your art, listen to constructive criticism. The end goal is to become a house-hold name.

- **Setting Prices**

One of the challenges many beginners face is not knowing how to set competitive prices. Beginners either set their prices too low hoping to attract buyers, or set them too high hoping for a quick windfall. Neither of these strate-gies work. As much as you want to set competitive prices, you must also be realistic about your work.

For example, as a beginner artist, people don't really know your work, so it might be a while until you are able to set premium-level prices. This shouldn't deter you, however. Research the market to understand what other artists charge for similar work, then price yours accordingly.

On realistic pricing, you should also factor in the cost of gas fees or minting. OpenSea has a gasless minting option. With gas fees costing upwards of $50 for an artwork, it's a bit prohibitive for beginners. Gasless minting is an alterna-tive solution that allows you to create your NFT on the blockchain without actually submitting it to the distributed ledger. The fact that you don't have to pay upfront for your NFT to be created will definitely attract more users to NFT communities.

If you decide to pay for gas, you must factor the cost into your price to make a decent profit. As a first-time artist, you probably haven't done so much to your work or brand in terms of promotion, so you'll need to be patient as it might be a while until you make a few sales. It might take you sometime to sell your art, but if you do it right and position your brand well, you might be making sales for a long time.

- **Build a Collection**

For a beginner, it might be easier to sell a collection of art than a single piece. Collections generally tell a story, or provide more context to a theme. Collections also work in your favor because if someone likes it, they'll buy the entire set so they have the complete story.

Presenting NFT collections is easier said than done. It's not just about bundling a few items together and putting a price tag on them. Put some thought into it. Remember that you're not just doing this for yourself, you are creating a story that your audience can relate to.

Other than the artwork, your collection should be marketable. This is about positioning your work to appeal to an audience. More importantly, you are targeting a community of artists, so you must position your work as such. Think about exposure while building your collection around a community. You can either use organic marketing or paid advertising. Engage audiences on social media. If possible, use all social media outlets you can access. Ultimately, your success or failure will depend on how you engage the community around which your art project is created.

Engaging your community frequently is the best way to market your collection, build hype around your work and create more awareness for your brand. The reward for positive engagement is user-generated awareness and traffic, and you don't have to pay anything for that.

Finally, create a compelling story around your collection. Visual storytelling is one of the most powerful tools artists use to engage audiences and build a loyal following. It adds a touch of purpose, personality, and emotion to your work. Create room for interaction by allowing users to comment, like, share and ask questions about your work.

- **Metaverse Exhibition**

Now that you've figured your way around NFTs, your next logical step should be the metaverse. Somnium Space and Decentraland are great starting points. Art metaverses are a good opportunity to promote your art and meet other amazing artists. There's a lot of growth in this space, making it one of the best places to pitch your work.

The idea of experiencing the real world through a virtual mirror pushes the limits of shared experiences between the physical and digital worlds of art. Fine art company Sotheby's recently launched their metaverse, bridging the gap between the traditional world of high-end art, and NFTs for the same. With major art outlets setting up a platform for digital collectors, this is proof that you need to get your work onto the metaverse too. Sotheby's alone has made more than $70 million in NFT art sales in an environment that's continually expanding to capture the world of digital arts.

The beauty of it all is that it's now easier to assign value to artwork through authentication and a history of exhibition. Many art collectors in different metaverses display their art collections on their galleries, virtual walls, or within their estates on Decentraland.

Art exhibitions in the metaverse are great for you because of the wide variety and experience you will encounter online. Besides, you don't necessarily have to limit yourself to the visual arts. You can create art out of anything, from 3D illustrations to lines of programming code. Everyone who contributes to the evolution of the metaverse is an artist in their own capacity, making virtual experiences a reality. The art world is already on the move. By creating NFT art, you are already playing a big role in the journey of merging digital technology with contemporary art.

Be keen on the level of activity in the network before you mint NFTs. This will help you plan around gas fees. Generally, minting when the blockchain isn't crowded is the best way to keep your gas fees low. It's all about demand and supply. You can use tools like Etherscan to monitor gas prices at different times of the day.

Finally, learn how to use testnets. These are programs that allow you to experiment with your idea without affecting the main blockchain. Rarible and OpenSea have this functionality, so you can experiment before minting the final artwork. They are quite useful for beginners who have yet to figure their way around the blockchain, so you don't write mistakes to the blockchain, and spend more to add correct entries to the blockchain.

CREATING VALUE IN THE COMMUNITY

Beyond the hype, one of the reasons why NFT spaces have experienced tremendous growth is because of the aspect of communities. You too can leverage this and use NFT communities to build a thriving business online. The two most important things in this approach is branding and marketing your work. If you do things right, this should make it easier to position your work in a manner that appeals to many audiences.

Blockchain is such a wide ecosystem with myriads of uses, some of which we haven't discovered yet. NFTs are but a subset of this ecosystem, whose popularity has been on the rise. One of the reasons behind this surge in popularity is because of the communal approach that developers have been pursuing.

People are working together to create amazing experiences through NFT communities. Learning opportunities, collaboration, and socialization are some of the factors that are driving more NFT developers to build their projects around communities. The good news is that it works, but only if you do it right.

Maybe one of the best examples of why it's important to create NFT projects around communities is in the metaverse. At the moment, the most we know about the metaverse is that there will not be one, but several. We also have an idea of some of the big tech brands that are building the metaverse, or lending their technologies to support the infrastructural growth in that sector. Laying the infrastructure for the metaverse is the first step. The most important one, however, is getting people to find value in the metaverse, and that's where communities come in.

Performing artists, for example, can hold concerts in the metaverse. Without a community of fans excited about this approach, it would fail. Therein lies the need for engagement, branding, marketing, and everything else you can do to create awareness of your brand. So, as we proceed with our community discussion, remember that there's more to it than just getting people together. It's about leveraging the numbers to boost your brand, and more importantly, providing tangible value to the members.

LEVERAGING SOCIAL MEDIA

The blockchain understanding of a community means a group of fans, users, enthusiasts, investors, developers, leaders, and anyone else who can contribute to the advancement of an idea on the blockchain in whichever capacity. Online communities mostly thrive on social networks like Discord, Twitter, and Instagram. This is where most of the discussions and engagements take place.

While all social media outlets are great for advancing your community, some might not necessarily be ideal for your

NFT project. This is why it helps to research well before you choose a social media outlet to advance the community aspect of your NFT project. The best way to approach this is by considering the kind of engagement you need. For example, interactive sessions are more productive on Reddit, where communities can ask and answer questions. You'll get lots of good feedback about your project right there.

If you wish to release some news about your NFT project, Bitcointalk is the most ideal platform to use. Bitcointalk is great for creating a buzz around breaking news and any other news-worthy information about NFTs.

Telegram and Discord are chill spots. People casually hang out on these platforms, and at times engage one another in in-depth discussions about different NFT projects. Twitter, on the other hand, is perfect for releasing information about progress on developing a project. It's also a good place to make your project trend and get more people interested in it.

Communities help you increase support for your project, learn and improve your credibility, and establish a good foundation on which you can promote your NFT projects in the future. While you can mostly engage audiences on different social media platforms on your own, you might need to consider hiring a marketing professional to help you navigate each of the social media platforms and maximize on their potential.

BUILDING AN NFT COMMUNITY

A strong and supportive community will help you advance your project faster, seeing that you'll have access to both fans and investors. This is why it helps to work on building your community from the onset. Let's briefly discuss useful tips to get you going:

- **Establish an audience**

Audiences have different needs, similar to the explanation we had for using different social media outlets. You must first figure out the specific needs of your audience, then build your community around that. The end goal here is to create value for your audiences, so start by understanding their needs.

- **Visualize the end product**

What's your vision for the community? The vision is actually as important as the NFT project. It will help you map the way forward, plan the scope of your community, and also help you figure out the kind of idea your community should buy into.

As you think of the vision for your project, you should also consider appropriate channels that will help you achieve it. Your use of social media will go a long way in helping you achieve this. Research what approach other artists are using for similar projects so you can get a good idea of the best approach to consider for your project.

- **Grow your community**

A community is built on engagements. Get people talking; create content that they can share across different platforms. Use all avenues available to encourage more activity in and outside the community you create. Remember that there's more to growing the community than just creating a few posts and hoping your audiences will make them trend. Building a community is an everyday task. Think of it like learning—it never stops. Be consistent in creating new informative pieces about your NFT project. The more people learn about it, the more they will challenge you and help your growth, both as an artist and at a personal level.

- **Building trust**

Communities thrive not just because of the commonality of shared objectives, but also because of the bond of trust. When you ask people to share your content within their wider networks, you are essentially asking them to put their faith in you as they encourage their friends, family members, and associates to buy into your project. Indirectly, you are asking them to be your brand ambassadors.

For all that you ask of your community, the best way to repay their kindness is by trust. Build a brand people can trust. Be a person people can trust. There's no doubt that competition in digital art NFTs is fierce at the moment. You are even facing competition from NFTs that are not related to art in any way. Amidst all this, your transparency and authenticity will appeal to more members of your community.

People have used underhand tactics in the past to try and gain some mileage in the NFT business. Some members of your community have fallen victim to such projects before, so rest assured they know how to spot a disingenuous project from afar. By all means, never try to sell an unrealistic idea to your audience. This will only breed failure. Besides, people have access to various tools and resources that can help them with due diligence.

One way to approach the trust issue is by issuing certificates of authenticity. This goes a long way in assuring collectors that you are the real deal. Anything you can do to allay fears of the legitimacy of your NFT project becomes a value addition to your community.

You must also strive to provide tangible benefits to the community. A good example of this is entrepreneur Gary Vaynerchuk's (Gary Vee) Flyfish Club. Gary Vee made headlines by promoting the first NFT restaurant in the world. This is the exclusive members-only restaurant we mentioned in an earlier chapter, and membership is open to anyone who owns the token. This means that you can sell your token, as many are already doing on NFT marketplaces.

The restaurant has two tokens, one allowing you access to the cocktail lounge and restaurant, while the other gives you access to both of them, plus their exclusive Omakase room. Now that's the kind of tangibility of benefits that would entice a community to buy into your NFT project.

For more context, Gary Vee's restaurant NFT raised more than $14 million in a week.

Communities are incredibly important in your journey to building a successful NFT project. In the next chapter, we'll cast our gaze beyond the present ecosystem, and see how NFT communities can help you position your brand for future success.

CHAPTER 8
CRYSTAL BALL TIME - THE FUTURE OF NFTS

Predicting the future isn't the easiest thing to do, especially in the realm of technology. For example, there was a time when Yahoo! was the go-to email platform for everyone. No one could have imagined that some years down the line, Google would take over not just the search engine business from them, but also the email business and a host of other services.

Apart from timing being the key element of a successful venture (just think of how MySpace failed massively despite a premature initial offering not too dissimilar to Facebook), there are other unknowns we must take into account when looking into our crystal ball and analyzing previous failures and uncertainties that bogged down companies like Yahoo and MySpace.

These uncertainties are also mirrored in the blockchain world. It's hard to tell what will happen in a year from now. However, we can look forward to exciting new developments in this space. It doesn't get more exciting than an influencer selling her love as an NFT for $250,000. Someone actually paid that much money for love! Marta Rentel, the Polish influencer behind this NFT, also tokenized her YouTube videos and Instagram content. Who would have thought that a time would come when we could put such a precise price on affection?

One thing that's certain about NFTs is that we will come across a lot of "first-time" news headlines, as people figure out innovative ways to monetize pretty much anything. If someone can sell love as an NFT, there really isn't a limit to what we can put a price on.

Another interesting discussion that we'll have to delve into as a society is setting clear boundaries between the real physical world and the online world. For example, the influencer who sold her love NFT was quite categorical on that matter. She stressed the fact that the internet is a virtual place and nothing on it can be physical. She further implored her fans and followers not to mistake her online persona from her real persona.

This is an interesting aspect which has also come up several times in discussions around the metaverse. The metaverse, at full capacity, should create an immersive experience that connects the real world to the virtual world. Architects, for example, could collaborate on projects in real time from anywhere in the world. This is possible by installing sensors at strategic points of the physical structure they are collaborating on, then creating 3D renders of the project in a collaborative space on the metaverse.

It's not just architects; we also have the wider creative industry, content creators, real estate, the healthcare sector, you name it. There's something in the works for everyone. Microsoft, for example, is primed to help the corporate sector get in on the metaverse craze. They've spent decades improving different enterprise-level solutions, so we can anticipate they will be at the forefront, pushing metaverse solutions into enterprise-level products. This should further speed up the adoption of the metaverse, seeing that many people will have first-hand experience at work, and gradually introduce the metaverse into their homes as the gadgets and tech become more affordable and easily available.

All these integrations happening around us gives credence to the love NFT influencer's point that we must be able to separate the real world from the virtual world. Tokenizing an emotion is a unique concept, but it won't be the last. Beeple crawled so other digital artists could walk. We'll most certainly see other people tokenizing emotions online.

While the jury might be out on the idea of tokenizing emotions, stranger things have happened in the creative community. From the Italian artist Salvatore Garau who sold an invisible sculpture to someone buying Twitter co-founder Jack Dorsey's original tweet as an NFT, the relationship between the traditional arts and the digital world will get more interesting with advancement in blockchain technology.

A THRIVING COMMUNITY

2021 was about the hype. The problem with hype is that it fizzles out after a while as people move on to other things. However, the NFT space is different because the hype brought forth the realization of utility, hence value in digital assets other than the currency approach. Amidst the hype, people realized that there's actually so much they can do with NFTs and get tangible value.

Future success for NFTs will be modeled around the one thing that's held humanity together for generations— community. That sense of belonging makes a difference in people's lives every day. NFT developers are now exploring the horizons beyond the hype, and the best way to do this is to establish communities around their projects. This is driven by the need to create long-term, sustainable benefits in NFT projects, and establish trust within those communities. This approach could also be an important step in creating new opportunities for collaboration and more importantly, advancing a new dimension for learning.

After the 2021 boom when people spent crazy amounts of money in NFT marketplaces, the market naturally settled into a lull. The early birds cashed out of some projects, probably because they felt that they were riding a trend that had run its course. Soon after, the agency or brokerage culture set in, where people would buy NFTs and flip them for a profit. None of this is strange, it is just who we are as a society. This is another manifestation of the concept of scarcity that we discussed earlier in the book, and how we create perceived value in rudimentary things because of their rarity. Left to these factors alone, the NFT market would have probably fizzled out and become a fad. However, building NFT projects around communities has been a game-changer. We are social animals, after all.

The allure of communities is also driven by the desire to be among the early adopters of NFT projects. If someone would have told you to invest $5,000 in Bitcoin at the early stages, you'd have laughed off that idea and thought they were crazy. In hindsight, that might have been the best idea you'd ever implemented had you taken up the offer. This trajectory has been witnessed in many blockchain projects, which start off as something of a joke, then they trend and go viral online, and before long, a community emerges, and the project becomes an exclusive community.

Developers of projects like the Doge Fight Club realize there's a lot of value in appealing to our sense of belonging. Other than that, there's also the promise of anticipatory rewards in the future. Buyers aren't clamoring for avatars anymore. Other than the gaming niche, the avatar ship has sailed. People are buying into NFT projects for the promise of entertainment, community, information, and education. This, perhaps, is also the right time for such advancements, as we inch closer to the metaverse.

There's a lot of insight in NFT communities, which is something many people seek. You get to learn so much about events in the crypto market, projects on the horizon, and what to add to your watchlist. You can think of this as the evolution of online forums. You'll come across discussions on everything from upcoming ICOs to voting on community DAOs.

The 2018 film *Ready Player One* presented a good example of what we might expect in a virtual world. If we consider everything that's been said about an immersive digital world, and that it might erode the nature of human interaction, it's quite ironic that advancement in the NFT spaces is being built around the concept of communities. We are already experiencing glimpses of the metaverse through multiplayer games like Fortnite. Fortnite is synonymous with weapons, fighting, and all kinds of violence. However, there's also a weapon-free version, Party Royale, where you can simply socialize. Fortnite temporarily removed the mode in April 2022, possibly to revamp it and bring back a better, more intriguing version.

Meta, formerly Facebook, has also been working on a project that's quite similar to the virtual reality depicted in Ready Player One, Facebook Horizon. This is also why they invested billions of dollars in the 2014 acquisition of Oculus Rift. As early as 2014, the company was already foreseeing an immersive future with virtual experiences, and the role of virtual reality in it. The Facebook Horizon project is supposed to encourage exploration and adventure, which also reiterates the new approach by NFT developers to create communities where people can interact and learn.

We'll see more actionable education content in NFT communities like how-to videos, tutorials, webinars and other content to further research and advancement in this space. Away from developers, users are also approaching NFTs differently. People have learned from others' risks, successes, and failures, and have higher expectations of NFT projects thanks to wide research and due diligence.

There was a time when ownership of most NFT projects was shrouded in mystery and anonymity, which resulted in a lot of fraudulent activities, scams and rug pulls. Some projects also collapsed simply because the founders lost interest or got busy with other things. These are some of the issues that come up in community discussions within NFT projects. Regulators have also cast their glance on NFTs, since they are increasingly becoming a part of the financial system. Through skepticism, they missed the train on cryptocurrency, and have mostly been playing catch-up. However, we can expect that regulators will be right there at the onset for most NFT projects we'll interact with in the future. With regulators in the fold, full transparency might soon become mandatory for start-ups and other projects in this space, giving users confidence in the projects.

A COMPARATIVE ANALYSIS

Other than the financial sector, the impact of NFTs has been greatly felt in the traditional art industry. There are notable similarities between the traditional art industry and the digital art industry, especially in terms of valuation.

The four important factors that help in determining the value of any work of art: demand, liquidity, market data, and an intermediary, are also reflected in the digital space.

Demand is created by the market, and often is all about the creator's status. There are currently many artists in the market, and it's quite difficult for new or unknown artists to sell their work at high prices. An artist's reputation, therefore, determines whether they command significant demand in the market or not.

Liquidity, relevant to demand, is about the resale value of the art. If you buy the work of a famous artist, it's easier to resell it later at a profit. This is where auction houses and art dealerships make their money. We also have famous art collectors who command a lot of respect and influence in the market. Such a collector could easily turn an unknown artist's fortunes around, because as soon as they buy their work, people will want to know the artist. You can think of this as celebrity endorsement.

The best example of demand in the digital art scene is Beeple, a digital artist who wasn't getting much attention for his work, but is now revered in NFT marketplaces. On the subject of liquidity, it's no secret that there's a growing market for digital works of art. Many art enthusiasts who have often felt priced out of buying quality art from the traditional market now have access to a world of art, and many at affordable prices. From rare digital works to simple pieces, you are always a few clicks away from an artist.

Through DeFi, artists now have leverage in that they can present their works as collateral for peer-to-peer loans on the blockchain. Thousands of dollars worth of loans have already been issued through this system, proof that there's a ready market for this business model.

Valuation also comes down to the data available in the market. Market research is paramount in this respect. It helps in appraising art, considering factors like the creator's identity, their style of work, equipment or material they use in their work, and so on.

These factors aside, both the traditional and digital art market is mostly speculative, and valuation will always be subjective. The price of a piece of art comes down to the perception of a few individuals in the market, powerful enough to decide whether something is valuable or not. This gatekeeper mentality is one of the reasons why many artists struggle to get their breakthrough in the traditional art scene.

The directness of blockchain technology means that the gatekeeper approach and subsequent role of middlemen in the industry might be a thing of the past. Middlemen have been eliminated from most commercial situations where they've profited over the years, including art.

Artists can now create and sell their work on their own terms, since they have direct access to audiences. They have more control over commissions, royalties and other forms of income from their work from both primary and secondary sales, which is quite a significant shift from the mode of operation in the traditional art markets.

Market data is just as relevant in the digital art markets as it is in the traditional art markets. If anything, it's even more important in the digital world. This space has unfortunately been host to some unscrupulous activities in the past that cost unsuspecting buyers and investors a lot of money. Unfortunately, the benefits of blockchain technology are not limited to users of pure intention. Criminals are also able to leverage the technology for their personal gains.

Possible solutions include strict governance standards and possibly introducing fees, which might deter some unscrupulous dealers, but still not all of them. Therefore, to protect your investment, it's important to perform due diligence before investing in any form of digital art. On a wider scale, you also have autonomous blockchain validation working for you.

Ultimately, both the traditional and digital art markets share similarities in terms of asset valuation. The dynamics of the factors might be different, but what remains clear so far is that valuation in the collective art market is a speculative matter.

Tremendous growth in the digital arts will continue into the future, but at what cost? We are already facing the obvious challenge of electricity consumption by blockchain assets. If the annual power consumption of Bitcoin alone ranks so high as to be included in the top 30 or 40 nations in the world by power consumption (mining for blockchain currently consumes as much energy as the country of Serbia), we need to take a step back and find feasible solutions right away.

The irony is that the environmental degradation as a result of blockchain transactions is real, yet all this drain on natural resources is to power structures of a virtual world.

Ethereum is taking real steps in this direction, by transferring their blockchain to a Proof of Stake verification model, which uses far less computational power, and thus consumes much less power.

THE FUTURE OF FINANCE

The global financial markets have thrived mostly because of collaborative regulation all over the world. In its most rudimentary element, blockchain technology and watch-dogs don't go hand in hand, especially in the banking sector. Yet, we could see a bit of regulation creeping into the blockchain as we advance it into the future of finance, and to encourage adaptability.

The good news is that this is disruptive technology, and every forward-thinking financial institution is looking at how to align themselves with these changes. Bringing blockchain technology into the financial fold will be a huge win for banks, especially since they already have the confidence and trust of their customers. These are two of the most important factors that make many users skeptical and rethink investing in the blockchain. However, banks can offer blanket immunity given their stringent regulatory protocols, making it easier for even the least tech-savvy of customers to come onboard.

Whether cryptocurrency will be the future of money, or if we will completely move away from the concept of money we know, is a matter of speculation and time. El Salvador tried it by making Bitcoin legal tender. The government gave everyone in the country free bitcoin to kickstart the economy - and it got mixed reactions. Some people were delighted with the free money, whilst some small business owners are worried about what a sudden drop in price to this volatile currency would do to their livelihood.

Within this effort in El Salvador lie important lessons on implementation, and more importantly, legislation on regulation. But the bottom line is that it can be done.

The million dollar question is, will every country eventually adopt cryptocurrency as legal tender? And if so, when?

Some of the pertinent issues as we march towards this eventuality include the need to identify and classify digital assets into the right asset classes, their acceptability as collateral, and their adaptability. These issues are critical in light of the nature of financial contracts and debt security.

Once El Salvador recognized Bitcoin as legal currency, financial authorities all over the world cast their attention towards this innovation, hoping to learn from their approach to integration, implementation, regulation, legislation, and more importantly, their mistakes.

One of the reasons El Salvador leveraged blockchain technology was to reduce the cost of remittances from citizens living and working abroad, eliminating capital gains tax on Bitcoin, which added value for residents holding Bitcoin. Adopting cryptocurrency could also signal the end, or at the very least, a seismic shift in the global dependence on the U.S. Dollar. Countries that depend on the U.S. Dollar generally have no control over their local monetary system, and will always be vulnerable to the impact of a struggling U.S. economy, if and when that happens. Accepting Bitcoin as legal tender also champions financial inclusivity by bringing more people into the financial system who would have otherwise been unbanked or underbanked.

Unfortunately, accepting Bitcoin as legal tender also comes with unique challenges. For example, merchants who might not be interested in cryptocurrency must unwillingly come onboard so that they don't miss out on business opportunities.

Volatility in Bitcoin prices is another challenge that could spell doom for widespread adoption of Bitcoin as legal tender. Imagine the inflationary pressure when a country holds its currency reserves in Bitcoin, then the price slumps, say from $40,000 to $30,000 or even lower. This would create a lot of economic anxiety and panic.

The success of cryptocurrency as legal currency will greatly depend on whether governments can offer the same level of legal protection and comfort as other financial assets enjoy. This possibly means that popular cryptocurrency exchanges will have to register, and conduct more due diligence in the interactions with their customers.

It's quite unfortunate that the global banking industry has mostly held onto the COBOL technology that's been around since 1959, being fiercely impervious and resistant to technological disruption—talk about fear of the unknown! Yet, we are on the cusp of a disruptive force so massive that the global financial sector is now forced to rethink their stand on technological integration, robustness of legislation on privacy, and more importantly, rethinking the structure of governance in central banks.

Even though banks and the global financial system at large must adapt to the changing times, they must approach the next step cautiously. It is a step in the right direction for economic development, innovation and financial inclusion, but also a step worthy of extreme caution.

THE NEXT STEP

The NFT market has so far shown the kind of resilience we've often experienced in stock exchanges and other financial markets. We've seen periods of heightened activity in the market and times when activities crashed and burned down so much that people started speculating as to whether the bubble had indeed burst.

These are normal economic events in every market. Our attention, however, should be drawn to the magnitude of the activities in the NFT market, as therein lie the answers to what the future holds for NFTs. According to a Forbes publication, the total NFT sales in 2021 was $25 billion, up from $95 million in 2020.

$25 billion worth of hype certainly doesn't sound realistic. Of course, there's been a lot of hype in the market, and many people invested their money in NFTs to get in on the next big thing. However, there are tangible reasons to believe in the sustainability of this sector. Case in point, the metaverse.

While the internet has come a long way, the next step in its evolution will be the game changer. The third generation of the internet, Web 3.0 is about creating an intelligent, and interconnected online experience for users. This will involve a lot of machine-level understanding of data. The blockchain ecosystem is awash with technologies that will help in making Web 3.0 a reality, from NFTs to cryptocurrencies, dApps, and smart contracts. We can already see their potential implementation in metaverse proposals by different brands championing the cause.

Isolating the NFT marketplace from other blockchain constructs, we see an industry that is growing tremendously on different levels. The art sector has always thrived on diversity, so we can expect further growth in that sector. Beeple might have grabbed the headlines with the insane amounts his artwork fetched at auctions, but this isn't the end. There will be more artists grabbing the headlines in the future.

Gaming is another area that's growing tremendously. Gaming and innovation go hand in hand. This also explains why the gaming sector is widely anticipated to be direct beneficiaries of the metaverse concept. Gadgets like virtual reality headsets that will be key to advancing the metaverse are already pivotal in creating enhanced and immersive experiences for gamers. Therefore, the transition from gaming into the metaverse should be smooth. Economically, gaming NFTs contributed close to 20% of all NFT sales in 2021, which proves the practicality of NFTs, and allays fears that people might have of NFTs being nothing but hype. As long as we keep investing in advancing the arts, gaming, entertainment, and other sectors aligned with NFTs, the NFT market will experience tremendous growth in the future.

The transformative nature of blockchain technology isn't just being experienced in other allied sectors, but also within the blockchain ecosystem itself. One area where this is vastly imminent and necessary is in the use of consensus algorithms. Bitcoin, and many of the altcoins that were built off the Bitcoin blockchain use the Proof of Work (PoW) consensus. PoW is notoriously heavy on resource consumption, and one of the reasons why blockchain technology has drawn a lot of criticism from environmentalists.

Consensus algorithms like PoW ensure that new blocks are added to the blockchain, verified and validated. It is also through these algorithms that order is maintained in the blockchain, hence the seamless function without the need for a central administration to oversee everything. Unfortunately, PoW consumes a vast amount of energy.

There are other consensus algorithms that have been considered by blockchain projects, which are not as resource-intensive, hence they are friendlier to the environment. Some examples are Proof of Burn (PoB), Proof of Stake (PoS), Proof of Capacity (PoC), and Proof of Authority (PoA). The Ethereum blockchain, for example, is moving from Proof of Work to Proof of Stake, which simply allows users to validate blockchain activities based on the amount of coins they can stake on the blockchain. At the time of this publication, the eligibility criteria for PoS on the Ethereum blockchain is 32 Ether.

Even though the alternatives to PoW are promising a reduction in the energy consumption used by blockchain activities, there will still be room for further development and refining the consensus algorithms. Ultimately, NFTs, cryptocurrencies, and everything else under the blockchain umbrella are widely advancing technologies, so we can be certain we haven't seen the last of the reinventions yet.

CHAPTER 9
CONCLUSION

NFTs are more than a step in the technological advancement of our society, they are a phenomenon that any future-minded individual should learn about. From the onset, we learned the practicality of NFTs in terms of their versatility. This is one of the reasons why it's easy to elicit NFT utility in every industry. The blockchain concept might be a tech idea, but its application transcends the tech industry.

While NFTs might gain mainstream attention through hype projects, that phase is coming to an end, and we are entering the age of value. Users across different industries are realizing there's so much potential in NFT projects, if they are implemented in the right way. It gets even better, in that NFTs are but a subset of what we can expect of blockchain technology. As much as there are many noteworthy NFT projects making headlines today, we are yet to experience the full potential of NFTs, let alone blockchain technology.

It's one thing to learn about NFTs, and another altogether to put that knowledge to use. My hope is that you implement the knowledge gained from this book towards creating a better future for yourself and those around you. I insist on including those around you because that is how you play an active role in advancing new technology that will have a lasting impact on future generations.

We are the people around us. The communities we belong to make us who we are. I learned about NFTs the difficult way, and decided to make my contribution to society by sharing my knowledge. This way, you get to follow a simpler, clearer route to understanding NFTs.

In terms of utility, NFTs have paved the way for artists from all walks of life to take bold steps in reshaping their future. If you are an artist, some of the concepts of the traditional art world won't apply to you anymore. NFTs bring you closer to your audience without wasting time and money on brokers.

With more advancements in the NFT ecosystem, we've also realized that it's possible to push the boundaries of art beyond its traditional descriptions. We can no longer limit ourselves to art in the form of pencil work, canvas paintings, sculptures, and so on. Art is bigger than that. The digital ecosystem has made it easier to expand the horizons of art. Even programming code can now be presented as an art NFT. The same applies to video games and other kinds of proprietary information.

Following the growth trajectory of NFTs, it's clear to see that there are endless opportunities in this field, and a lot of room for everyone to figure out how to stake their claim in the future of the internet. The metaverse is the future; NFTs are just the vehicles to usher us into it. There's also the fact that virtual environments give you the power to create your own reality. You can create things that might not be feasible in the physical world.

As much as NFTs are drawing all the attention, we must not ignore the reality of their impact on our environment. Most NFTs are currently running on the Ethereum platform, whose annual energy consumption is as high as the consumption of some industrial nations. We are burning non-renewable sources of energy to power an emerging technology whose value is mostly virtual. At some point, we have to take a step back and rethink our priorities.

The Ethereum blockchain, for example, is transitioning from Proof of Work to Proof of Stake. While this change in consensus mechanisms is welcome, it might be a while until we can discern the real magnitude of change. One problem we might encounter is that as much as we might be able to reduce the energy consumption by some degree, there will be considerably more people using blockchain technology in various capacities.

Over time, NFTs and other blockchain projects will be adopted in different sectors as more people realize their inherent benefits. Central banking institutions might also follow the example of El Salvador and start integrating blockchain solutions into local financial systems in some capacity. Therefore, even if Ethereum and other blockchains move to Proof of Stake and or any other alternative consensus mechanism, the massive demand for blockchain technology might still pose the same environmental threat that we are facing at the moment. We just don't know the magnitude yet.

Overall, the most important lesson you should take from this is the utility value in NFTs and blockchain technology in general. There are many opportunities you can tap into. From creating your personal brand of NFT, to implementing blockchain-level solutions to solve real-life problems, there's so much you can do with this technology.

This book represents a moment of pride for me, knowing that you have all the tools you need to get started in the NFT ecosystem. Like I mentioned at the beginning, it took me two years to gain this knowledge, learning the hard way. You don't have to go through that. I trust that you now feel more comfortable engaging in discussions about NFTs, cryptocurrencies, the metaverse, and any other blockchain subject that might come up, and are more assured to invest in NFTs with the confidence of a seasoned veteran investor.

You are now equipped with the basics to dive into elementary investing in cryptocurrency, and if you're an artist, you now have the tools to create your own presence on the blockchain, so go out there and enjoy it!

CHAPTER 10
A QUICK FAVOR

Reviews are the bedrock of my success. Taking 5 minutes of your time to review this book will benefit me in one of two ways. If the book wasn't to your satisfaction, please leave constructive criticism to make me a better author. If you enjoyed the book, a good review will give my book more clout in the Amazon algorithms and generate more exposure for my books. I'd be extremely grateful if you could rate my book now. Thank you.

CHAPTER 11
REFERENCES

Adar, D. (2021, November 29). *The underlying motivations of NFT trading - UX Collective.* Medium; UX Collective. https://uxdesign.cc/the-underlying-motivations-of-nft-trading-504e4036dda7

Bhowmick, S. (2021, October 31). *How To Build An NFT Community? Learn From The Founders Of NFT Malayali.* Https://Www.outlookindia.com/; Outlook India. https://www.outlookindia.com/website/story/how-to-build-an-nft-community-learn-from-the-founders-of-nft-malayali/399392

Big Think. (2021). Bitcoin and blockchain 101: Why the future will be decentralized | Big Think [YouTube Video]. In *YouTube.* https://www.youtube.com/watch?v=eHCOuqrZx18

Birnbaum, J. (2022, January 14). Why Video Game Makers See Huge Potential In Blockchain—And Why Problems Loom For Their New NFTs. *Forbes.* https://

www.forbes.com/sites/justinbirnbaum/2022/01/06/why-video-game-makers-see-huge-potential-in-blockchain-and-why-problems-loom-for-their-new-nfts/?sh=7fcbf55043d7

Brain, L. (2021, September 24). *Top 5 Real-World Non Fungible Tokens Use Cases - DataDrivenInvestor*. Medium; DataDrivenInvestor. https://medium.datadriveninvestor.com/top-5-real-world-non-fungible-tokens-use-cases-946d92591fbf

Casey, M. J. (2021, September 3). *The Value of NFTs Is Belonging*. @Coindesk; CoinDesk. https://www.coindesk.com/tech/2021/09/03/the-value-of-nfts-is-belonging/

Cass, J. (2022, February 27). *10+ Different Types of NFTs - Complete List*. JUST™ Creative. https://justcreative.com/types-of-nfts/

Cho, R. (2021, September 20). *Bitcoin's Impacts on Climate and the Environment*. State of the Planet. https://news.climate.columbia.edu/2021/09/20/bitcoins-impacts-on-climate-and-the-environment/

Clark, M. (2021, March 3). *NFTs, explained: what they are, and why they're suddenly worth millions*. The Verge; The Verge. https://www.theverge.com/22310188/nft-explainer-what-is-blockchain-crypto-art-faq

Coinbase. (2022). *What is cryptocurrency?* @Coinbase; Coinbase. https://www.coinbase.com/learn/crypto-basics/what-is-cryptocurrency

Connor, D. (2022, February). *Advantage and Disadvantages of El Salvador Making Bitcoin Legal Tender*. Blockchain Files. https://blockchainfiles.org/advantage-

and-disadvantages-of-el-salvador-making-bitcoin-legal-tender/

Conti, R. (2022, May 13). What Is An NFT? Non-Fungible Tokens Explained. *Forbes*. https://www.forbes.com/uk/advisor/investing/cryptocurrency/nft-non-fungible-token/

Criddle, C. (2021, February 10). *Bitcoin consumes "more electricity than Argentina."* BBC News; BBC News. https://www.bbc.com/news/technology-56012952

DappRadar. (2022). *NFT (Non-Fungible Tokens) | DappRadar*. DappRadar. https://dappradar.com/nft

Dartroom. (2022). *Dartroom*. Dartroom.xyz. https://dartroom.xyz/overview

Digital Asset News. (2019). What is a Non Fungible Token NFT vs. a Fungible Token? [YouTube Video]. In *YouTube*. https://www.youtube.com/watch?v=gbsewBbW3Jk

DiTullio, J. (2021, February 25). *NBA Top Shot: What is it and Why is it Popular?* The Game Haus; The Game Haus. https://thegamehaus.com/nba/nba-top-shot-what-is-it-and-why-is-it-popular/2021/02/25/

Dunn, S. (2021, March 9). *Unstoppable Domains Has New Record For Most Expensive Domain Name NFT*. Crypto-daily.co.uk; CryptoDaily. https://cryptodaily.co.uk/2021/03/Unstoppable-Domains-Has-New-Record-For-Most-Expensive-Domain-Name-NFT

Dunn, S. (2021, May 24). *Sorare: The NFT Brand Aiming to Become NBA Top Shot for Soccer - Boardroom*. Boardroom.   https://boardroom.tv/sorare-nft-crypto-fantasy-football-cards/

Ennetht. (2022). *Community – A Key Metric For NFT Investing*. Ennetht.com. https://www.ennetht.com/all-about-nft-blog/nft-deep-dive/community-a-key-metric-for-nft-investing

Fanusie, I. (2018). *NFTs bring live event tickets "to life," YellowHeart CEO says*. Yahoo.com. https://uk.news.yahoo.com/nf-ts-bring-live-event-tickets-to-life-yellow-heart-ceo-says-131654265.html

Fatemi, F. (2022, April 14). Here's How NFTs Could Define The Future Of Music. *Forbes*. https://www.forbes.com/sites/falonfatemi/2022/01/24/nfts-and-the-future-of-music/?sh=1990761b5677

Fayre Labs. (2022, March 4). *How To Build Your NFT Community - Fayre - Medium*. Medium; Fayre. https://medium.com/fayrelabs/how-to-build-your-nft-community-7b9fafaa7175

Flyfish Club. (2021). *Flyfish Club | Home*. Flyfish Club. https://www.flyfishclub.com/

Flynn, Dr. T. (2021, April 29). *The Provenance Research Blog*. The Provenance Research Blog. https://www.theprovenanceresearchblog.com/home/2021/4/29/mgx6kfmk0fd1j3gssbbna5v8f0ka7y

Footprint Analytics. (2022, February 16). *How to Choose the Right NFT Marketplace? | CryptoSlate*. CryptoSlate. https://cryptoslate.com/how-to-choose-the-right-nft-marketplace/

Fries, T. (2019, December 11). *CryptoKicks: Nike to Tokenize Shoe Ownership on Ethereum*. The Tokenist. https://tokenist.com/cryptokicks-nike-to-tokenize-shoe-ownership-on-ethereum/

Georgiev, G. (2022, January 24). *Top 10 Most Expensive NFTs Ever Sold (Updated 2022)*. CryptoPotato. https://cryptopotato.com/most-expensive-nfts-sold/

GET Protocol. (2016). *The Future of Ticketing | GET Protocol*. GET Protocol: The Future of Digital Ticketing. https://www.get-protocol.io/

Gomezz, A. W. (2021, April 24). *Cyber Scrilla*. Cyber-scrilla.com. https://cyberscrilla.com/how-non-fungible-tokens-are-changing-sports-forever/

Gomezz, A. W. (2021, November 10). *Cyber Scrilla*. Cyberscrilla.com. https://cyberscrilla.com/top-12-most-valuable-nft-trading-cards-available-on-the-market/

Harper, J. (2021, March 23). *Jack Dorsey's first ever tweet sells for $2.9m*. BBC News; BBC News. https://www.bbc.com/news/business-56492358

Harris, C., & Thomas, L. (2022, April). *The Top NFT Music Moments of All Time*. Nft Now; 351Studios. https://nftnow.com/music/top-music-nft-moments/

Harris, J. (2021). NFTs, Explained [YouTube Video]. In *YouTube*. https://www.youtube.com/watch?v=Oz9zw7-_vhM&t=3s

Hawkins, J. (2022, January 13). *NFTs, an overblown speculative bubble inflated by pop culture and crypto mania*. The Conversation. https://theconversation.com/nfts-an-overblown-speculative-bubble-inflated-by-pop-culture-and-crypto-mania-174462

Hayes, A. (2022). *10 Important Cryptocurrencies Other Than Bitcoin*. Investopedia. https://www.investopedia.com/tech/most-important-cryptocurrencies-other-than-bitcoin/

Hellon, M. (2021). *Stagwell | What Does Psychology Tell us About the Explosion of NFTs?* Stagwellglobal.com. https://www.stagwellglobal.com/what-does-psychology-tell-us-about-the-explosion-of-nfts/

Howcroft, E. (2022, January 11). *NFT sales hit $25 billion in 2021, but growth shows signs of slowing.* Reuters; Reuters. https://www.reuters.com/markets/europe/nft-sales-hit-25-billion-2021-growth-shows-signs-slowing-2022-01-10/

Hunnewell, D. (2022, April 20). *The 20 Best NFT Projects to Follow in 2022*. Coffee Bros.; Coffee Bros. https://coffeebros.com/blogs/nft/the-20-best-nft-projects-to-follow-in-2022

Ivelina. (2021, April 17). *How to Become a CryptoArtist in 10 Steps - NFT Plazas*. NFT Plazas; NFT Plazas. https://nftplazas.com/how-to-become-a-cryptoartist/

Katje, C. (2018). *Why NFT Sneakers Could Be Coming Soon From Nike*. Yahoo.com. https://finance.yahoo.com/news/why-nft-sneakers-could-coming-181755561.html?guccounter=1&guce_referrer=aHR0cHM6Ly93d3cuZ29vZ2xlLmNvbS8&guce_referrer_sig=AQAAAIm-Gn96C7G9H-S9Yf0_ecLVJgXKEQcqoC8ACgreegDUJ2Rz4HELaB7a2

bCtznsM9FnloZddMtOlALDC7i5x25IBPmK3oFrKmJQh
CdjcZseHJrhcEwZYhNhbx1inPxqbBfBflVyfxubuxl3_vIpI
TVQPlpT0t32vIgo6TJjP-0z1

Kiger, P. J. (2021, May 17). *Cryptocurrency Has a Huge Negative Impact on Climate Change*. HowStuffWorks; HowStuffWorks. https://science.howstuffworks.com/environmental/conservation/issues/cryptocurrency-climate-change-news.htm

Knibbs, K. (2022, February 8). *How Did the Bored Ape Yacht Club Get So Popular?* Wired; WIRED. https://www.wired.com/story/celebrity-nfts/

Lielacher, A. (2020, December 5). *"Traditional" Art vs. Crypto Art: How to Value It*. Cryptonews.com; Cryptonews. https://cryptonews.com/exclusives/traditional-art-vs-crypto-art-how-to-value-it-8527.htm

Lyons, K. (2021, March 22). *Jack Dorsey's first tweet sold as an NFT for an oddly specific $2,915,835.47*. The Verge; The Verge. https://www.theverge.com/2021/3/22/22344937/jack-dorsey-nft-sold-first-tweet-ethereum-cryptocurrency-twitter

Marcin, T. (2021, June 20). *Classic memes that have sold as NFTs*. Mashable; Mashable. https://mashable.com/article/classic-memes-sold-nft-prices

Matt's Crypto. (2021). How to Buy and Sell NFTs For Profit (Full EASY Beginner Guide) [YouTube Video]. In *YouTube*. https://www.youtube.com/watch?v=XFfzjWCO-SU

Michael, A. (2022, May 13). What Is Cryptocurrency? *Forbes*. https://www.forbes.com/uk/advisor/investing/cryptocurrency/what-is-cryptocurrency/

Monty, S. (2022, January 6). *A Brief History of NFTs: Where Do They Come From and Where Are They Going?* NFT News Today. https://nftnewstoday.com/2022/01/06/a-brief-history-of-nfts-where-do-they-come-from-and-where-are-they-going/

NFT Explained. (2021, October 5). *NFTExplained.info*. NftExplained.info. https://nftexplained.info/what-is-meant-by-community-in-an-nft-project-complete-guide/

Pennington, A. (2021, June 8). *The Metaverse: Where Did It Come From and Where Is It Going?* NAB Amplify; NAB Amplify. https://amplify.nabshow.com/articles/ok-so-how-exactly-do-you-metaverse/

Perkmann, M. (2021, October 23). *Is the NFT Bubble About To Burst? - Coinmonks - Medium*. Medium; Coin-monks. https://medium.com/coinmonks/is-the-nft-bubble-about-to-burst-d8158ea3f75a

Prescott, T. 'Jett'. (2021, June 4). *How NFTs Are Redefining Our Cultural Mentality Around Value*. Rolling Stone; Rolling Stone. https://www.rollingstone.com/culture-council/articles/nfts-redefining-cultural-value-1173699/

Quiroz-Gutierrez, M. (2022, February 7). *Nike is the latest company to file suit over NFTs as brands get serious about the marketplace*. Fortune; Fortune. https://fortune.com/2022/02/07/nike-files-nft-lawsuit/

Reese, F. (2019, April 29). *7 Benefits of Decentralized Currency*. Bitcoin Market Journal. https://www.bitcoinmarketjournal.com/decentralized-currency/

Shahaab, A., & Khan, I. (2020, October 7). *Estonia is a "digital republic" – what that means and why it may be everyone's future*. The Conversation. https://theconversation.com/estonia-is-a-digital-republic-what-that-means-and-why-it-may-be-everyones-future-145485

Sharma, R. (2022). *Cryptokitties Are Still a Thing. Here's Why*. Investopedia. https://www.investopedia.com/news/cryptokitties-are-still-thing-heres-why/

Simplilearn. (2019). Blockchain In 7 Minutes | What Is Blockchain | Blockchain Explained|How Blockchain Works|Simplilearn [YouTube Video]. In *YouTube*. https://www.youtube.com/watch?v=yubzJw0uiE4

Sorrentino, F. (2022, April 21). Banking On Blockchain. *Forbes*. https://www.forbes.com/sites/franksorrentino/2021/11/29/banking-on-blockchain/?sh=685035e370ed

Sothebys. (2022). *Sothebys*. Sothebys.com. https://www.sothebys.com/en/search?query=nft&tab=auctions&cmp=actn_CTP_gg_sea_sar___en_6-21___ser_searesp___nft&s_kwcid=AL%2113028%213%21564783132647%21p%21%21g%21%21nft%20artworks&gclid=CjwKCAiAvaGRBhBlEiwAiY-yMAYva7A2WhDvu6qJ_LvPt3SPvSszDgmjvhhBkXFIJMOHwrCh7JplVRoCyWkQAvD_BwE

Steinwold, A. (2019, October 7). *The History of Non-Fungible Tokens (NFTs) - Andrew Steinwold - Medium*. Medium; Medium. https://medium.com/@Andrew.

Steinwold/the-history-of-non-fungible-tokens-nfts-f362ca57ae10

Syed, A. (2021, July 28). *An influencer sold her love as an NFT for $250,000 and is going to have dinner with the mystery buyer*. Business Insider; Business Insider India. https://www.businessinsider.in/thelife/news/an-influencer-sold-her-love-as-an-nft-for-250000-and-is-going-to-have-dinner-with-the-mystery-buyer/articleshow/84854846.cms

Taylor, E. (2021, November 15). *The Next Wave of NFTs is Starting With Community First*. Decrypt; Decrypt. https://decrypt.co/85808/the-next-wave-of-nfts-is-starting-with-community-first

Tetrina, K., & Schmidt, J. (2022, May 17). Top 10 Cryptocurrencies In May 2022. *Forbes*. https://www.forbes.com/uk/advisor/investing/cryptocurrency/top-10-cryptocurrencies/

Thaddeus-Johns, J. (2022). Beeple Brings Crypto to Christie's (Published 2021). *The New York Times*. https://www.nytimes.com/2021/02/24/arts/design/christies-beeple-nft.html

Ticketpark. (2021, December 14). *NFT tickets — A Realistic Look At A Big Trend - Ticketpark - Medium*. Medium; Medium. https://medium.com/@ticketpark/nft-tickets-a-realistic-look-at-a-big-trend-ae813d6f885d

Token Minds Editorial Team. (2021, August 13). *NFT Crypto Community: 6 powerful strategies of building an engaged society for your project*. TokenMinds; TokenMinds. https://tokenminds.co/blog/knowledge-base/nft-crypto-community-how-to-build-nfts-community-for-success/

Tyfield, S. (2022). *Cryptocurrency – The Future of Banking?* Shoosmiths.co.uk. https://www.shoosmiths.co.uk/insights/articles/cryptocurrency-the-future-of-banking

Vaswani, K. (2017, February 13). *What is blockchain and how does it work?* BBC News. https://www.bbc.com/news/av/business-38932854

Walkden, E. (2021). *How to be a successful NFT artist: Advice from 4 artists*. Verisart. https://verisart.com/blog/post/how-to-be-a-successful-nft-artist

Waterworth, K. (2021, December 6). *Investing in NFT Real Estate*. The Motley Fool; The Motley Fool. https://www.fool.com/investing/stock-market/market-sectors/financials/non-fungible-tokens/nft-real-estate/

Zwieglinska, Z. (2021, October 4). *How fashion brands are navigating NFTs and what's next for the metaverse*. Glossy; Glossy. https://www.glossy.co/fashion/how-fashion-brands-are-navigating-nfts-and-whats-next-for-the-metaverse/

www.ingramcontent.com/pod-product-compliance
Lightning Source LLC
Chambersburg PA
CBHW071654210326
41597CB00017B/2202